the colour of innovation

Made in Brunel
Brunel University, Uxbridge, UB8 3PH
+44 (0) 1895 267 776
www.madeinbrunel.com

Art direction by Nick Sardar.

Editors:	Product Photos:
Olly Brown	Dave Branfield
Heather Bybee	Patrick Quayle
Madeleine Carver	
Clive Gee	General Photos:
Stephen Green	Olly Brown
Kirsti Macqueen	Ross Dexter
Samuel McClellan	Richard Harris
Joe Midgley	Dominic Sebastian
Paul Turnock	Max Woźniak
Emma Tuttlebury	
Max Woźniak	

Type set in Franklin Gothic and Fedra Serif.

First published in 2011 in collaboration with
Papadakis Publisher.

An imprint of New Architecture Group Ltd. Head
Office: Kimber Studio, Winterbourne, Newbury,
Berkshire, RG20 8AN. www.papadakis.net

ISBN: 978-1-906506-15-5

Butler Tanner & Dennis

This book is printed using 100% vegetable based inks on Respecta Silk FSC.
Respecta Silk FSC is produced from 100% Elemental Chlorine Free (ECF)
pulp that is fully recyclable. It has a Forest Stewardship Council (FSC)
certification and is produced by a mill which supports well managed
forestry schemes. The whole production process is managed on one site
to further reduce the impact on the environment.

Foreword

James Caan, CEO, Hamilton Bradshaw

Business is built on the vision of innovative thinkers. It is people that make great businesses and I have always been passionate about encouraging and advising people who have the rare gift of being business-oriented innovators. Made in Brunel is the incubation platform for a new generation of innovative thinkers.

I am a great believer in promoting the very best of British science, technology and innovation in the UK and Brunel exemplifies this. It is one of our finest universities, with a philosophy of developing creative and practical ideas with commercial objectives.

"Made in Brunel is the incubation platform for a new generation of innovative thinkers."

Made in Brunel is a unique project that represents inspired, educated and motivated people; three words that should underpin all pioneering ventures. I wish this project every success and I am sure that the talented young people behind Made in Brunel will move into successful business careers and go on to help shape so many aspects of our future in this country.

James Caan, CEO, Hamilton Bradshaw

a

Abdel-Gadir, Mohamed	140		
Abidi, Murtaza	343		
Acquaye, Richard	327		
Al-Twal, Nadim	191		
Albano, Daniela	322		
Alder, Lucy	112		
Amar, Ashima	172		
Ambridge, Alexander	206	259	
Amjadi, Nazanin	194		
Anderson, Dave	258		
Arding, Steven	312		
Arthur, Kylie	176	191	193
Atefipour, Kourosh	265	279	
Auriol, Antoine	325		

b

Babatola, Seun	98	285	
Bamford, Penny	231		
Barbalata, Sorana	58	213	
Bennet, Julian	228		
Blakeney, Stephen	229		
Boggia, Jade	110	282	
Branco-Rhodes, Yinka	92	142	
Bright, Edmund	297		
Brockie, Simon	214		
Brooks, Ashley	180	192	194
Brown, Olly	51	123	
Brunel Masters Motorsport	302		
Brunel Racing BR12	300		
Bruno, Elena	212		
Bybee, Heather	184	188	193

c

Carver, Madeleine	158	224	253
Ceren Tan, Fatos	182		
Cervantes, Andres	181	342	
Chaleampao, Nopadon	344		
Cham, Jason	264	288	
Chan, Nathan	56	298	
Chang, Yao-Wen	241		
Chien, Yu-Hsin	241		
Chung, Grace Minjoo	194		
Chung, Michael	256	266	
Clarke, James	236	292	
Clarke, Timothy	297		
Claydon, Dean	347		
Cole, David	138		
Collett, Tom	148	242	
Connor, Mark	101	254	
Couch, William	307		
Cullen, Fiona	323		

d

Daly, Michael	315		
Dawes, Laurence	335		
Day, Michael	152	286	
Denchie, Shirley	349		
Devoy, Hannah	40	274	
Dexter, Ross	50	237	
Dhillon, Jugbir	315		
Dhillon, Onkar	316		
di Palma, Amelia	139		
Diaz del Castillo, Juan	304		
Dodoo, Ansbert	346		
Dong, Haoqiong	190		
Dunkley, Timothy	117	162	
Durler, Carl	326		
Dy, Goldbert	316		

e

East, Robin	315	
Edgar, Nicholas	25	165
Ellson, Tom	296	
Elsouri, Mohammed	26	153

f

Fahimi-Rad, Farzam	316	
Fisher, Dee	76	94
Fletcher, Jon	44	62
Fraser, Sebastian	325	342
Frederico, Luiza	185	188

g

Gana, Grace	191		
Garcia, Abraham	295		
Ge, Qing	190		
Gramon-Suba, Iulia	175	192	194
Grannum, Seán	338		
Gutteridge, Steven	63	339	

h

Haile, Hannah	100	164	
Hall, Claire	183	191	194
Hamid, Muhammad	315		
Hamiliton, Victoria	33		
Harris, Richard	23	275	
Hassan, Turgay	116	318	
Hassell, Katie	228		
Henbest, Katie	39	122	
Ho, Kai Yeung	315		
Hodder, Adrian	108	257	
Hope, Emmanuel	284		
Horan, Nicholas	52	194	342
Hsu, Szu-Chuan	241		

j

Jackson, Nicholas	312	
Jay, Tom	204	255
Jeganathan, Victor	272	280
Jiang, Shaochen	241	
Jones, Philip	226	
Jones, Tom	249	

k

Kaewkanjai, Noppan	230	
Kan, Carson	225	
Kemp, Joel	66	215
Kierans, Lucy	131	160
Kim, Mina	188	
Kirk, Benjamin	171	188
Kneller, Christopher	297	
Knight, Chris	306	
Köchert, Emanuel	65	269
Koren, Katy	72	157
Kwok, Sze Yin	225	

l

Latimer, Max	37	132	
Lee, Hyein	170	188	
Lee, Jiyun	241		
Lee, Sun	188		
Legg, Jermaine	161	223	
Li, Zheng	290		
Li, Yanan	192		
Liao, Yu-Ching	241		
Lowther, Allan	273		
Lun, Zhen	82		
Luyken, Roman	111	268	340
Lyons, Hugh	342	345	

m

Macqueen, Kirsti	24	31
Makwana, Pritesh	33	

Malhotra, Rohan	270	342		
Malik, Fawad	309			
Malone, Michael	336			
Maltby, Thomas	267	281		
Martinez Rodriguez, Maria Paula	177	194		
McAvoy, Rebecca	214			
McClellan, Samuel	45	239		
McCormack, Jennifer	173	192		
Meng, Yupeng	231			
Mensah-Ansong, Efua	99	243		
Midgley, Joe	27	216	238	
Millard, Daniel	296			
Mistry, Premal	156			
Moir, Unji	233			
Monk, Nicholas	312			
Mudannayake, Jude	313			

n

Nicholls, Stephen	42	

o

Odell, Rohin	97	200
Oprisan, Andra	174	192

p

Palmano, Buster	220	248
Palmer, Joseph	229	
Park, Haeyoon	193	210
Parrish, Matthew	95	201
Patel, Rishee	330	
Pearse, Adenike	191	
Pharez, Charlotte	43	124
Piper, Antony	33	
Poon, Michelle	80	113
Puttock, Michael	86	289

r

Rahimi, Keahn	334	
Rahman, Mohammed	228	
Ramnath, Anthony	324	
Reader, Tom	126	293
Rezaei Zadeh, Javad	316	
Rezaei Zadeh, Reza	316	
Rojas Monserratte, Carlos	191	
Rose, Rebeckah	283	
Rowinski, Matthew	87	287
Rugg, Jonathan	89	277
Rutter, Theo	296	
Ryan, Calum	333	

s

Sadiq, Shabaz	228	
Saeed Qureshi, Mohammed Umair	309	
Sandén, Tor	211	
Sardar, Nick	28	73
Schneider, Jenny	53	64
Schofield, George	308	
Sebastian, Dominic	30	41
Sekar, Sathiya	228	
Shabani, Arber	156	342
Shadbolt, Thomas	229	
Shakespeare, James	331	
Shanker, Tobias	33	
Shim, Hyunhee	189	
Sidhu, Inderjit	214	
Simko, Michal	305	
Simpson, James	229	
Singh, Anita	214	
Smith, Adam	48	74
Smith, Daniel	130	227
Smith, Jonathan	309	

Smith, Matt	90	109	
Song, Eunchung	189		
Sook Han, Jung	192		
Stanley, Michael	205	348	
Strickland, Christopher	22	102	251
Subramony, Saravana	228		
Suerdas, Sarp	344		
Suh, Min	193	194	
Swallow, Fred	163	202	
Swayne, Trevor	332		

t

Tashie-Lewis, Bernard	317		
Thirumurthy, Ashwin	342	343	
Thomas, Philip	203	252	
Thompson, Ross	294		
Thwaites, Toby	312		
Trevena, Simon	309		
Trigg, Jamie	91	151	
Trigkaki, Konstantina	189		
Tsai, Ming-Chih	210		
Tsikolis, Theodore	178	189	194
Tulloch, Jayson	141	299	
Tuttlebury, Emma	107	271	
Tzarela, Dimitra	189		

v

Van Olmen, Tilia	167	
Verma, Sam	150	207

w

Wade, Tom	211		
Wang, Bing	81	115	
Watts, Matthew	179		
Weerakone, Marshel	314		
Wells, Rosanna	32	125	
Wherry, Bradley	166	240	
White, Adam	297		
Whitehead, Benjamin	106	114	
Wickens, Stuart	133	246	
Wilkin, Daniel	276		
Williams, Harrison	221	250	328
Williams, Peter	36	222	
Williamson, Alex	93	149	
Willson, James	88	247	
Wong, Samuel	38	46	77
Woo, Sunhye	193		
Wooldridge, Rob	134		
Woźniak, Max	15	75	291

x

Xiang, Qiqi	190	
Xiao, Tingting	52	
Xue, Haoyun	345	

y

Yan, Meihui	83	
Yang, Ivan	194	230
Yang, Shuzhen	190	
Ye, Wei	190	
Youell, Charles	296	
Yu, Chi	337	
Yu, Xiaofan	190	

z

Zambrano Novoa, Johanna	189	
Zeitler, Philip	57	278
Zhang, Hongchen	181	
Zhang, Lu	232	

Isambard Kingdom Brunel

1806 - 1859

In the 100 years up to 1860, a small group of engineers led the way during the period of economic and social upheaval in Britain that we call the Industrial Revolution. Isambard Kingdom Brunel was perhaps the most exceptional of this group, and many of his works, which challenged and inspired his contemporaries, have survived to our own time.

Entering his father's business at the age of 20, Brunel's first project was to drive a tunnel under the Thames from Rotherhithe to Wapping – a complex feat of engineering. At the age of 26, he was appointed Engineer to the newly-formed Great Western Railway and his civil engineering works on the line between London and Bristol are still used by today's high-speed trains, demonstrating the quality and endurance of his designs. The project encompassed not just the rails, but also viaducts and stations, including Bristol Temple Meads, at the time the biggest railway station in the UK.

His other works included the Clifton Suspension Bridge over the River Avon, which was finished after Brunel's death in 1864 and is still in use today, as well as a remarkable prefabricated hospital, complete with air-conditioning and drainage systems, for use in the Crimean War. Brunel's three major ship-building projects, the Great Western, Great Britain and Great Eastern, launched between 1837 and 1859, represented major steps forward in naval architecture.

As his sketchbooks and notebooks show, Brunel concerned himself with every aspect of the projects in which he was involved, from the grace of a design to the precision of its execution. His great achievement was to marry his vision of how things could be done better with the calculations and experiments required to make it possible. He faced inevitable setbacks and disappointments, but he readily admitted his mistakes and indeed often suffered financially through supporting his ventures with his own money.

Brunel died of a stroke on 15 September 1859. Despite the short duration of his career, he is remembered as one of the most illustrious engineers of the Industrial Revolution, and his achievements still inspire designers and engineers the world over. His example is a particular inspiration to the staff and students of the university named in his honour.

> "Brunel was working 100 years before the emergence of the design profession, but was using design thinking to solve problems and create world changing innovations."
>
> Tim Brown, CEO, IDEO, TED Talk 2009

Brunel University

The history and ethos of Brunel University

Brunel is a world-class university based in Uxbridge, West London. Our distinctive approach has always been to combine academic rigour with the practical, entrepreneurial and imaginative approach pioneered by our namesake, Isambard Kingdom Brunel.

The University was founded in 1966 as a new kind of institution dedicated to providing research and teaching that could be applied to the needs of industry and society. This goal was central in the creation of our Royal Charter, one of the most forward-looking and visionary charters of its day. It emphasised the University's commitment to the relevance of academic learning, ensuring that teaching and research benefit both individuals and society at large.

The confidence and sense of purpose that characterise today's Brunel stem from this principle, which is still essential to our Mission, "to advance knowledge and understanding and provide society with confident, talented and versatile graduates", and vision, "to be a world-class creative community that is inspired to work, think and learn together to meet the challenges of the future".

Research is at the heart of all we do. We place great value on the usefulness of our research, which improves our understanding of the world around us and informs up-to-the-minute teaching as well as creating opportunities for collaborative work with business, industry and the public sector.

Moreover, Brunel's research ethos generates an atmosphere of innovation which inspires students and staff throughout the University, as well as encouraging the sharing and communication of ideas and expertise for which we are famous. Brunel's reputation is built on its groundbreaking work in subjects as diverse as design, engineering, education, science, sociology, IT, psychology, law and business. That reputation has gradually spread to new areas, including the performing arts, journalism, environmental science and sport and health; research and teaching in these areas still retains the same outward-looking philosophy.

A long succession of developments and mergers, particularly those which transformed Acton Technical College into Brunel University and which later saw Brunel merge with Shoreditch College and the West London Institute, have brought the University from modest beginnings to become a major force within the UK higher education sector and on the international stage.

Brunel's campus has been transformed over the last few years. The University's investment in

"Our distinctive approach has always been to combine academic rigour with the practical, entrepreneurial and imaginative approach pioneered by our namesake, Isambard Kingdom Brunel."

buildings and infrastructure is now approaching £300 million, with many new and refurbished social, teaching and sporting facilities and more campus green spaces.

We can now boast world-class sports facilities, a renovated Lecture Centre and Students' Union building, and an extended library with a hugely increased book and journal collection, more computer workstations and group study areas and an Assistive Technology Centre for disabled students. A new residential development, the Isambard Complex, has brought our campus accommodation up to 4,549 rooms in 34 halls of residence.

The Schools of Health Science and Social Care and Engineering and Design have benefitted from new buildings and facilities. A £30 million flagship building at the main entrance to the University is scheduled for completion in 2012, providing a new home for the Brunel Business School as well as an auditorium and art gallery.

Since the 1960s, Brunel University has provided high quality academic programmes which meet the needs of the real world and contribute in a practical way to progress in all walks of life. We are very proud of the University's development and of its contributions to the empowerment of individuals and to the progress of society. Most importantly, we have evolved a culture of innovative thinking that is applied, not just ideas in isolation but realised ideas. We apply our creativity and ingenuity across the broadest spectrum of subjects, working to add value and to build viable and exciting solutions to everyday challenges. We are future thinkers.

School of Engineering and Design

A centre of excellence

From 1966 and the origins of Brunel University; industrially and technologically focused research and education has been at the heart of what we do. The School of Engineering and Design is a globally recognised centre of excellence in design, engineering and innovation and continues as a flagship for innovation in the tradition of Isambard Kingdom Brunel.

The School offers 20 undergraduate courses and 20 postgraduate courses with a total student population of nearly 2500 divided between five subject areas. 340 PhD students from around the world work with leading researchers in an impressive array of specialisms, for example from biomedical engineering to environmentally cleaner electronics and inclusive design. All teaching and research is supported by extensive and diverse workshop and laboratory resources and dedicated technicians. This is the infrastructure providing the foundations to the output you can see within this Book.

A significant proportion of the work carried out within the School includes direct collaboration with commercial organisations, at all levels of study. In the earliest stages of the engineering and design students' University experience they will have completed projects with leading organisations such as TomTom, Bentley, British Airways, BAA, the BBC and M&S. Many undergraduate students undertake an industrial placement year and may continue their connection with projects in their final year. These links with industry enhance the real world relevance and currency of our work and ultimately lead to strong lifelong career prospects for our graduates. The alumni of the School form a global resource of connections which further enhance the potential for multidisciplinary collaboration and opportunities to tackle design and engineering challenges.

These are exciting times to be involved with design and engineering. There is increasing recognition of the importance of innovation and the role of our subjects to help tackle the major challenges of the 21st Century. The School of Engineering and Design is a formidable powerhouse of human and practical resources to embrace these opportunities.

Design

Home to the highest rated design courses in the UK (Sunday Times, 2011) and with postgraduate students from around the world, Brunel Design excels in technically sound human orientated design. We work extensively with industrial partners, applying innovative techniques to solving real problems. Over 80% of our undergraduate students complete a year in industry and we have an impeccable employment record. We specialise in design for the environment, branding, design management, inclusive design and design for manufacture.

Electronic and Computer Engineering

We produce engineers, technologists and designers for roles in a digitally connected society. Our industrial partners include Walt Disney, Microsoft, Avid, Xerox and Dare Digital. We manage over £5 million of research projects based around future developments in telecommunications and systems. 600 undergraduate and 300 postgraduate students are supported by 50 academic staff. Our specialisations include broadcast media design and technology, computer systems, Internet engineering, sustainable power through to electronic and electrical engineering.

Mechanical Engineering

Global challenges require engineering expertise. In the field of sustainability, mechanical engineering has had high profile success, being the most successful university entry in the electric motorcycle TTXGP event on the Isle of Man and the 'Water Cycle' human powered water purification device. We offer specialised engineering courses in aerospace, aviation, building services, automotive and motorsports. We are team workers, integrating mechanical engineering within projects including the highly successful Brunel Racing team.

Civil Engineering

We are addressing the important needs of the 21st Century, especially in sustainable building construction, roads, bridges, tunnels, flood protection, waste recycling and construction management. A core part of our ethos is sustainability, paralleling the 'cradle-to-grave' approach promoted by major national and international Engineering organisations. Our technical facility, the Joseph Bazalgette Laboratories, includes the latest technology to develop and test innovative materials. We are developing innovative solutions to civil engineering challenges.

Advanced Manufacturing and Enterprise Engineering

With focus on postgraduate education and research, this subject area develops world class expertise in advanced manufacturing technology, enterprise engineering and engineering management. The Advanced Manufacturing and Enterprise Engineering research laboratory has specialist facilities supporting: Micro/nano manufacturing/metrology, design of ultra-precision machines, robotics and manufacturing automation, 2D/3D vibration assisted machining and advanced manufacturing using lasers.

West London

A global hub for innovation

The quadrant of London sweeping out from Kensington and Chelsea, encompassing Wembley Stadium, Heathrow Airport and the Thames, M4 and M40 corridor travelling westwards is one of the most economically successful and vibrant regions in the world.

West London comprises of more than 750,000 jobs and 67,000 businesses, which annually contribute over £27 billion to the UK economy. Heathrow, the world's most successful airport, is a natural global hub for the world. The region has a tremendous track record as a home for innovative business at the cutting edge of global markets including the likes of GSK, BA, Disney, Coca Cola, Diagio, and BSkyB.

Joint research conducted in 2007/8 amongst large corporates and SMEs in West London and funded by the London Development Agency, clearly showed the high level of innovation achievement within West London whilst also identifying a keen appetite for sharing and gaining more knowledge on emerging innovation methods and techniques.

Collaborations between universities and business are increasingly recognised as important catalysts for innovation and economic regeneration. Many of the people and projects which make up Made in Brunel exemplify collaborations to harness the energy, creativity, research and knowledge available within the entrepreneurial spirit of the area.

The following organisations and initiatives are part of the infrastructure within West London to provide practical support to facilitate connections and the innovation which can flow from these:

West London Partnership

Chief executives and leaders of the six local authorities in West London and senior staff both from large corporate businesses and from SMEs based in West London join forces as WLP to develop the overall social and economic interests of the region. This dynamic partnership between business and local authorities has a broad remit including transport, regeneration, skills and workforce development, spatial development, planning and property, and housing.

West London Business

A business membership organisation representing a broad cross section of blue chips and SMEs across West London. West London Business provides a range of member services, while promoting economic development, entrepreneurial activity, exporting and innovation. It lobbies local, London and national authorities on behalf of the private sector, supporting the many growth sectors in West London, such as pharmaceuticals, food, IT, logistics, creative industries and tourism. West London Business is committed to raising economic competitiveness, creating new jobs and retaining businesses in the area, while promoting the principles of social inclusiveness and sustainability.
www.westlondon.com

Designplus

Designplus promotes design based collaborations between industry and universities. Based in Brunel University since 2004 Designplus establishes collaborative projects, leads events and provides professional development.
www.designplus.org.uk

© West London Business

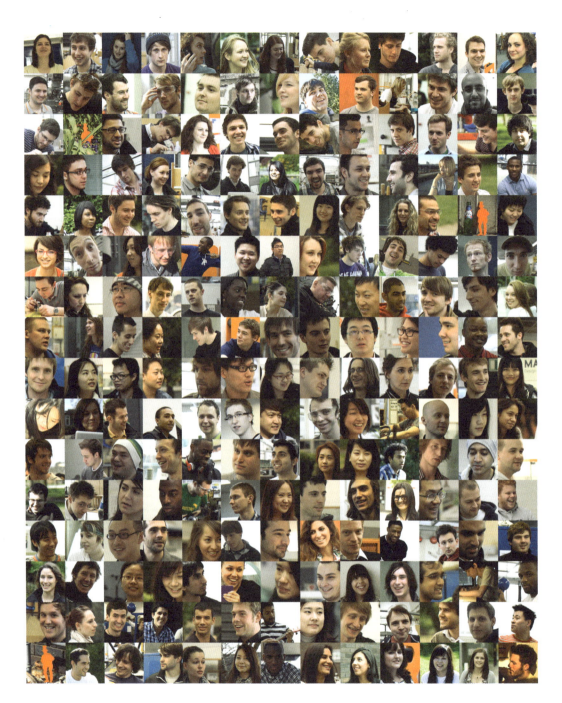

MADE IN BRUNEL™

Made in Brunel

In the past, spaces such as this have been used to wrap up a project, describing it as "finished". But for us, Made in Brunel is not a finished project. It is not a yearbook or a prospectus, nor is it an exhibition or degree show...it is our brand. Year in, year out, it continues to evolve while being carefully nurtured by hugely talented and passionate teams of people.

Our aim for 2011 is to portray the character and growth of these people through Made in Brunel and beyond. They begin their journeys in the professional world of Design and Engineering having grown into some of the UK's brightest innovators, and they really do hold the future of our world in their hands - it looks promising.

the colour of innovation

Our book represents a milestone near the very beginning. It captures not only the evolution of projects and their impressive outcomes, but the personality of our brand and our people.

Inside you will find stories from Made in Brunel ambassadors, sharing their journeys beyond this great chapter in their lives. They are testament to the calibre of students filling our book pages every year.

Made in Brunel sets itself apart from others, with every book carrying with it a new surprise, and each time gaining more presence. By the way... we do all of this in our spare time.

It's big, wonderfully exciting, bright orange... it's **the colour of innovation.**

Max Woźniak
Director 2011

contents

humanistic innovation 19

sustainable innovation 197

technical innovation 245

digital innovation 321

network directory 353

Products and concepts that improve the lives of people.

humanistic innovation   19

sustainable innovation   197

technical innovation   245

digital innovation   321

network directory   353

Future Concepts for Foursquare?

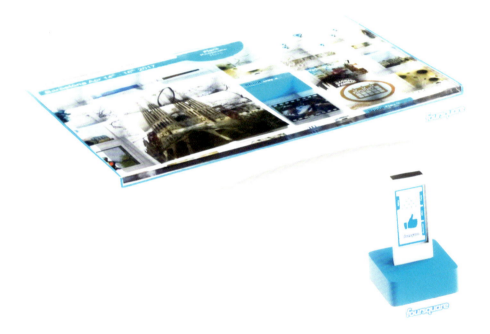

Foursquare is a location based social network, popular in America and spreading across the connected world. In 2020, Foursquare will be widespread and ready to release its first range of branded products. Ubiquitous computing has become the norm; people expect to be able to connect anywhere, with all of their social networking tools. Foursquare is more than just point-and-click broadcasting of arbitrary information to the masses; it allows people to engage with the real experiences of their closest friends. Foursquare products bring a deeper, more personal, interaction to the sharing of 'very-rich content'. By bringing the humanity back into everyday interactions we have with technology, together these products aim to make the benefits of social networking accessible to all across the broadest generational spectrum.

Foursquare Gaze

Future concept for Foursquare

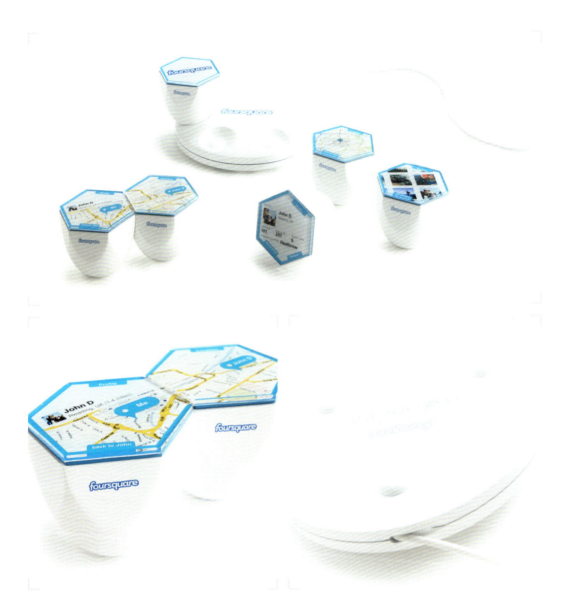

Foursquare Gaze allows those at home to follow the lives of those they care about most. By choosing which closest friends they want to keep close by, they get a richer interaction with the information that their friends share about daily lives. By physically playing with the Pods, people are able to combine different forms of content in a more interesting way to create a more engaging social networking experience.

Foursquare Gaze has two components; six Pods provide the interaction with up to the minute information and the Home which keeps the Pods charged and provides a safe place to store them. When each pod is taken away from Home, wireless connections ensure that they are kept up to date with the latest information. In this way the experience can be taken anywhere in the home.

Chris Strickland
Product Design Engineering

The compass is an item familiar from the childhoods of many. From exploring in the garden to field trips with the Scouts, the compass has been used in a number of different situations and is a key tool for navigation. The Foursquare Compass has been designed with the aim to provide the user with more information and feedback from their outing, all while maintaining a classic appearance. The concept not only displays a classic compass face but also has built in technology to provide the user with the following information: distance travelled, height above sea level, detailed log of your route, heart rate, calories burnt, navigation system. These can then be used to help the user to achieve new badges and awards with Foursquare.

Richard Harris

Industrial Design and Technology

Foursquare Keepsake Table

Future concept for Foursquare

Today, Foursquare is a fast-growing location based social network. In 2020, it will be global, and looking to extend its appeal to all age groups. By encouraging and supporting exploration, Foursquare will get all generations discovering their world, both local and further afield. Focusing on returning the physical experience to digital memories, the Foursquare Keepsake Table stores small souvenirs for reminiscence and sharing. Take a small object out of the box, place it on the glass top, and all of the associated images, videos, sounds and notes will appear on the screen for perusal. The large, intuitive touchscreen interface allows memories to be shared by young and old alike, while the lap-friendly unit encourages physical interaction with people and objects.

Kirsti Macqueen

Industrial Design and Technology

This device and docking system has been designed for the location based social network Foursquare. The check-in device allows people to carry it around with them at their own leisure, whether they are embarking on a trip or a simple visit to town. The aim is to personalise the 'checking-in' experience by providing an interaction between the user and a 'Foursquare hotspot'. These hotspots can include attractions and venues across the world, from the top of the Eiffel Tower to simply at your local grocery store. The user checks-in with the device by touching the LCD screen and flicking their thumb toward the check-in point, ultimately providing a more personal feel to location based social networking.

Nicholas Edgar
Industrial Design and Technology

3D Body Scanner Enclosure

In collaboration with Tailormade London

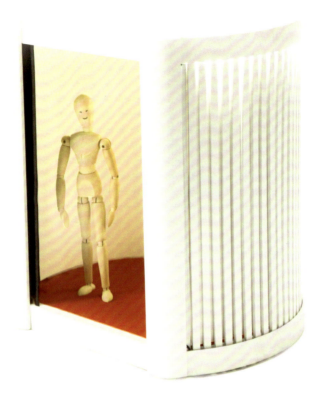

This project covered three particular challenges. Firstly, to enhance the user's experience inside the enclosure. The experience required design specifications essential to all types of changing environments such as clothes storage, seating space and privacy. Another aspect was the practical and technical function of the enclosure. The assembly and disassembly process needed to be simple, intuitive and user friendly. This area of the design defined the product architecture and covered several manufacturing aspects. Finally, the brand had to be integrated into the final product. This was about transferring the brand essence into the design of the enclosure both in form and function.

Mohammed Elsouri
Product Design

A reassuring and easy to use domestic fire extinguisher

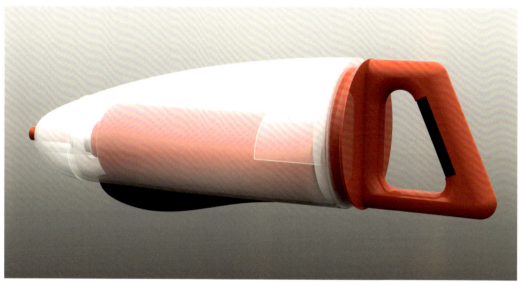

Many of us see fire extinguishers on a daily basis, but how many people actually know how to use them? This product focuses on fire safety in domestic situations, and the difficulties people face whilst using fire extinguishers. It looks at the process in a holistic manner and aims to reduce the amount of tasks required for successful operation. The extinguisher is holstered in a bracket, which encourages the user to remove the extinguisher by grasping the handle; a silicone grip on the main body promotes a firm and secure grip. The direction in which the powder is dispersed is indicated by the front nozzle and is achieved by pushing the rear handle. The transparency of the main body ensures the internal pressure vessel is visible, therefore ensuring the product retains fire extinguisher affordances.

Joe Midgley

Industrial Design and Technology

Aurora Control Console

Improving film production with reliable, rugged rental equipment

Much capital equipment in the film and television industry is functional at best, suffering from a lack of investment in behavioural research before design and production. When camera professionals want Motion Control, they want it working as soon as possible. Maintenance or set-up issues cost valuable time and money. Motion Control facilitates visual effects, empowering a production team to realise their creative visions.

Ethnographic research on a bustling film set and interviews with BBC cameramen shooting in the Arctic gave user empathy, resulting in key insights for innovation. Collaborating with Kontrol Freax Ltd, the rugged Aurora Console allows for precision movement of camera pan, tilt and roll through balanced hand wheels. Complex operations are controlled through an intuitive touch screen.

Nick Sardar
Product Design

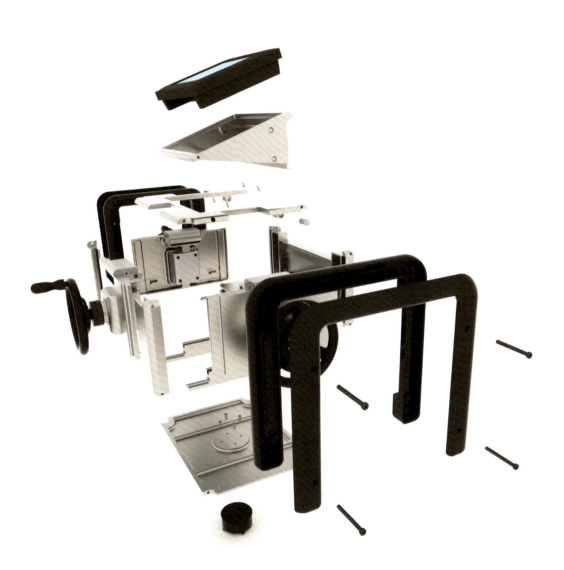

A Conceptual Rethink of a Neonatal Ventilator

An emotive approach to medical product design

The main issue in medical product design is balancing allocation of resources and efficiency. This feature led approach has long been seen in neonatal care, however during several visits to a range of leading London hospitals, it was found that an emotive approach to a redesign would be much valued by the nurses and parent users alike. In collaboration with SLE (specialists in Neonatal Ventilation for over 50 years) a new approach to the design of their medical equipment has been established. The concept makes use of a renewed product architecture, aiding cable management and reducing the existing ventilator footprint. Business goals underpin these design decisions, by integrating modularity and an external screen ensures a product with longevity, critical for the life span of their flagship product.

Dominic Sebastian
Product Design

Dementia affects one in five people over 80 in the UK. It is associated not only with loss of memory, but also loss of self-efficacy and self-esteem. By reminding an elderly user of the need to brush their teeth, then providing support as they negotiate the process, Cleo enables them to maintain independence and pride in their own care, while also allowing a carer to monitor their success. Consisting of an ergonomic manual toothbrush, a unique glass and an interactive base unit, Cleo guides the user through the process. A personal audio guide, recorded by a loved one, is complemented by text alerts and colour coding to help keep them on track. Styled to be approachable and elegant, Cleo represents an attractive, empowering assistive product.

Kirsti Macqueen

Industrial Design and Technology

Chameleon Shoes

Affordable bespoke high heels to suit anyone

High heels are a vital part of every outfit but buying a stylish and also comfortable pair of high street heels is a real challenge. Chameleon Shoes are a pair of high heels which have been designed around your foot so you know they will be pleasant to wear. To have a pair of hand-made shoes comes at a price and currently only comes in one style. This product allows the user to change the top section to suit different outfits, meaning they need only buy the upper part of the shoe each time, saving money and waste. Comfort comes also from foam insoles designed around the individual's foot to minimise the chance of discomfort. This product therefore not only has benefits for women who struggle to find a comfortable pair of shoes but also for different health conditions such as Diabetes as the insoles can be specifically designed around their needs.

Rosanna Wells
Product Design Engineering

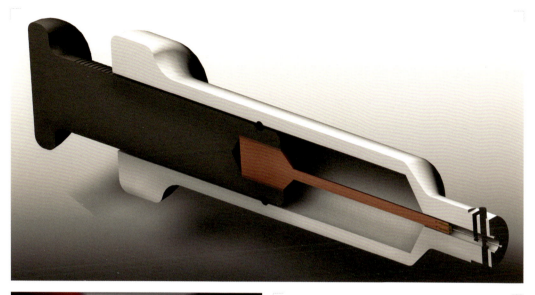

> "Studies showed that cement placed closer to the cortical layer greatly improved the pull-out strength of the screw."

Patients who have suffered fractures may require bone plates and screws in order to fully recover. Not every patient has the bone quality necessary for this, especially elderly patients with osteoporosis. The project aimed to develop a procedure of cement injection into the fracture site which would reinforce the screws in patients otherwise unable to receive them. Studies showed that cement placed closer to the cortical layer (outer layer of bone) greatly improved the pull-out strength of the screw. The design can determine the depth of the cortical layer and improves on the cement distribution within the bone. After extensive testing, the team managed to create a cannula which can deliver cement at the best location under the cortical layer in order to maximise pull-out strength of screws.

A. Piper, P. Makwana, T. Shanker, V. Hamiliton
Mechanical Engineering

James Hebbert

Made in Brunel. Batteries Not Included.

On arrival at the monastery deep in the south of England the ancient oak doors were bolted behind me. There had been an amusing misunderstanding: the monks had prepared me a chamber fit for a month's stay.

A design project that addressed sound control in public spaces led to a concept device that would effectively be the 'remote control to sound' and research was needed into the tangible qualities of silence. I unearthed an ancient religious order and interviewed a Carthusian monk on one of the rare days he was permitted to break his vow of silence. Discovering the values by which the monks lived was a gem for my project. This certainly was not the orthodox approach.

Determined to continue my unorthodox ways (as Brunel had always encouraged) to learn a very different attitude towards design, I set my sights on the Land of the Rising Sun. I jumped on a plane bound for Tokyo, not knowing where my adventure would take me.

A slice of raw bacon and grated cabbage welcomed me for breakfast at the Yamaha Corporation Design studio in Hamamatsu, where I began my first internship designing electronic musical instruments. I learnt a tremendous amount working alongside world-class designers and relating it to my own education. Take for example the definition of 'perfection': in the west, this is an ideal and so has a degree of unobtainabilty. By contrast, the Japanese understanding is of something that is 'complete' and, therefore, obtainable. Subtle nuances such as these can transform the outcome of a design.

A homebound plane awaited me had I been unable to find my next job so I resorted to persistent knocking on doors and in contrast worked for a small three-man design studio. Disillusioned by having to design the umpteenth USB key, I asked the senior designer, "Surely there can only be so many designs for the same product?" He wisely replied, "Think of design like cookery and creating a new food. You don't need to change it drastically, just add small details for it to taste better."

Like the USB key, I found myself repeatedly plugged into new environments gaining more knowledge each and every time. In some instances little English was spoken and I was forced to draw to communicate. I usually fail to tell people the number of companies that rejected me but during my time in Japan, I was fortunate to work for eight different organisations ranging from advertising to architecture.

Although once popular to inject fresh thinking, foreign internships became rare after the recession in the early 90's. A certain CEO of a corporation remarked that I had been the first western student to have cracked the Japanese internship system for over a decade.

Experience gained working abroad is invaluable; if you open your mind to a different cultural perspective it will bestow on you a unique selling point back home. I was fortunate to be selected as one of fifteen from 1,500 to be granted a marketing fellowship at the global advertising agency Ogilvy & Mather.

The ideals of the original 'Mad Men' are now extinct, as advertising is no longer a one-way communication. It should be a conversation that relates to the consumer emotionally. In this sense there is a strong parallel between advertising and product design since any product should radiate its core brand values.

After four years in the Ogilvy London office, I gained valuable experience working directly with the UK Chairman before being shipped across the pond and employed by the Global CEO in the USA.

"The future lies with the developing markets"

Currently, I work in an account management client servicing role in Ogilvy's New York office. The teams of 'account' and 'creative' are typically siloed in large agencies so to have an appreciation of the creative thinking, that Brunel gave me, combined with an understanding of the business needs helps me to extract the best from my team.

There is no other profession where I would have already become a specialist in industries ranging from spirits to shampoo, credit cards to chocolate. As I write, I have just returned from a shoot in LA for yoghurt, and I recently took a film crew around a selection of swanky Manhattan bars to ply people with alcohol for blind-tasting research on a vodka brand.

But really, it is not about the merchandise. I tell people that I am in the 'behavioural change' industry since I am intrigued by what makes people tick and how brands can build lasting relationships to add value away from the traditional techniques of advertising. The role that behavioural economics plays in creating intangible value from advertising is fascinating and I urge you to watch a presentation from Ogilvy's Rory Sutherland by visiting TED.com the instant you finish reading this.

Everything is driven by the digital muscle and the rate of innovation in technology is significantly influencing how brands talk to consumers, a shift also apparent in product design. Already we see technology dictating the way people behave to the extent that social etiquette towards technology has radically adjusted. We are creating for a world without wires: a mobile and socially connected planet. Take 'mobile' - a category now evolved to include tablets as well as phones - as an example: with an abundance of data at the marketeer's fingertips, the potential for brands to utilise the global mobile platform to engage consumers in a relevant, location-specific and timely manner, is phenomenal.

The future lies with the developing markets so my next chapter is to return to Asia. We can already see how 'Made in China' is evolving to 'Designed in China' as the credibility of Chinese brands stand strong in western markets.

Fusing the spectrum of skills from an engineer's technical mind with the right-side brain of an artist, Brunel Designers are a unique breed of trained 'solution finders', taught to think and execute under set constraints to add value to any creative industry and ready to meet life's challenges. But when you are Made in Brunel, the batteries are the only thing not included: it is the zest, perseverance and creative curiosity of a Brunel Designer that fuels the passion to succeed.

James Hebbert, Account Managment, Ogilvy & Mather
Graduated in 2005

Pro Grip

Improving the health and strength of ageing wrists, developing grip

The ageing process affects us all, and through this process muscle strength and our joint health begin to degrade. It is important to maintain the mobility and strength in our hands, as they allow us to carry out so many tasks and overcome unforeseen obstacles in everyday life. Hand exercise products on the market are demanding on the human body, not accommodating to the more fragile requirements of an elderly user.

Pro Grip is a hand exercise device, which helps develop one's grip and overall hand strength during the ageing process. The device focuses on comfort, offering greater user consideration in its design detail, material choice and usability. A simple internal mechanism allows the user to first develop their joints and when they are ready increase the muscle strength in their hand through adjustable spring resistance.

Peter Williams
Product Design

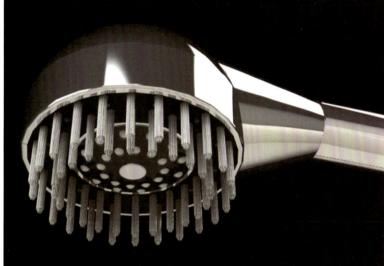

The market for products that empower people to maintain their quality of life and to enable independent living is increasing. Globally there are over 500 million people over 65 years old, a number that is predicted to exceed 835 million by 2025. Unfortunately, while age may bring increased wisdom it most certainly brings, sooner or later, a decrease in motor ability. The Inclusive Shower Handset is an assistive device that aims to make showering easier for all, but targets the older generation in particular. This handset enhances the bathing experience through good ergonomic, functional and emotional design to encourage feelings of self reliance and increased self esteem. High quality materials and attractive aesthetics add to the product's desirability and the kudos of owning such a useful and empowering device.

Max Latimer

Industrial Design and Technology

Tactile Canvas

Physical communication for the visually impaired

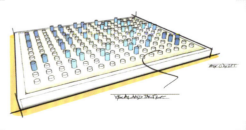

The canvas allows the manipulation of a matrix of contactors giving users the freeform ability to create tactile information. This approach brings to light new ways in which those with varying visual acuities can interact with each other. Braille messages, diagrams and games can be created and shared. Much quantitative research exists on why people do or do not use braille. This can be linked to education and levels of visual impairment, but only a moderate amount of qualitative research on the actual experience of people has been carried out. Working with these users revealed unexpected new areas of interest and needs of an invisible nature were uncovered. Braille is often misunderstood as a necessity for those who are visually impaired. This project was about understanding its capabilities and making them universal.

Samuel Wong
Product Design Engineering

Improving the experience of making hot drinks for the visually impaired

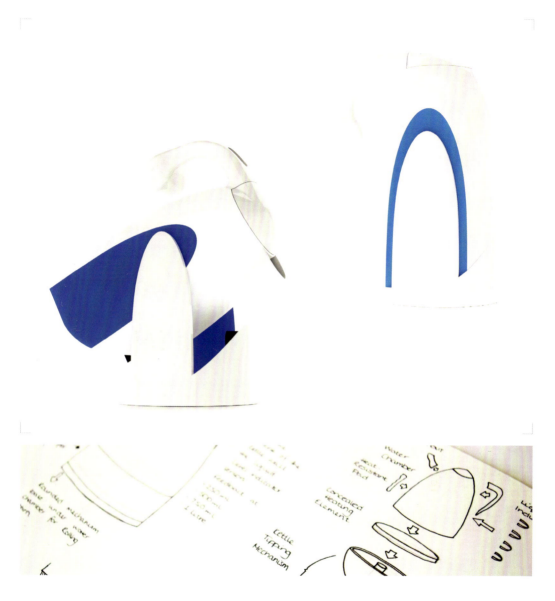

Currently, over 10 million people in the UK are aged over 60 years old, and this is expected to continue rising. A major implication of this ever-expanding 'ageing' population is the increase in age-related conditions, specifically visual impairments and dexterity issues. Making a hot drink is a vital part of the daily routines of many, but our ability to successfully carry out this task can be compromised as we age. Existing assistive products can reduce people's self-esteem and leave them feeling incompetent. Simpulo is a kettle, with an in-built tipping mechanism that allows the user to pour boiling water with ease, removing the strain upon the wrist. Additional vibratory feedback is provided to alert the user to the liquid level within the kettle. Simpulo carries no negative stigma and boosts the confidence of people in their own capabilities.

Katie Henbest
Industrial Design and Technology

Eddi

Future concept for Wonderbra

When times are hard and you need a friend, Eddi's glowing face will remind you of the people that care and give you the lift you need to succeed. Eddi charms are sold in pairs so that you can keep one for yourself and give its partner to a loved one. When both are worn at the same time, they light up and emit warmth to let you know that someone is thinking about you. Eddi connects wirelessly with your mobile phone through EddiApp to allow you to access a personal 'memory' that has been pre-stored onto Eddi by your friend or loved one. Eddi can be charged with the Wonderbra Charge Pad. Wonderbra Charm Bracelet sold separately.

Hannah Devoy

Industrial Design and Technology

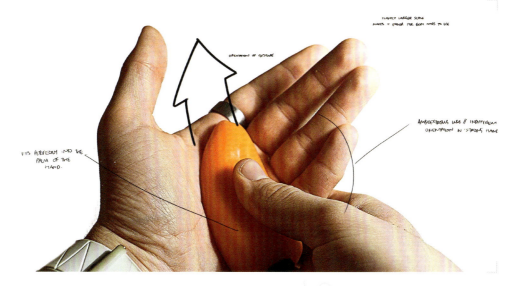

Menopausal symptoms have been found to be alleviated with a rich support network in place. Whether this network consists of friends or family, the value of the idiom "a problem shared is a problem halved" is very important for women in the autumn years of their life, where a shared experience can dilute what at first can be a very worrying or embarrassing situation for them. Taking that Wonderbra currently offer a very literal support to their buyers, it is only natural to develop the brand values to offer their buyers 'support for life', in this case to the age group of menopausal women. The concept is to build on the accessibility of social networks and media, to allow women going through menopause to connect with friends so menopausal moments can be turned into positive, shared experiences.

Dominic Sebastian
Product Design

Upper Body Posture for the Older Mouse User

Reducing musculoskeletal disorders in computer mouse use

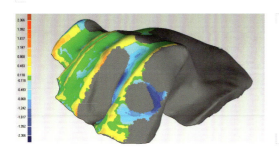

Posture for the older computer worker (or indeed for younger users) has been little addressed in the work environment. This device guides one's arm to a more central, and relaxed position when controlling the cursor and therefore allows the body to assume a more balanced posture. Combined with an ergonomic hand and wrist position, it allows the hand and fingers particularly, to rest when not working.

This creates less tension through the upper body and reduces the chance of Musculoskeletal Disorders. The device also provides ergonomic cursor control to the 20% of the population that are left-handed. Guided by research the design can be used in the left and right hand for alternate periods to give the most comfortable working experience.

Stephen Nicholls
Product Design Engineering

Arthritis can reduce motor capability in people's hands, resulting in difficulty performing basic tasks in the kitchen, such as using a kettle. Redesigning 'domestic water processing' to remove exclusion due to physical impairments should not carry stigmatism. In fact, an 'inclusive' product can carry usability benefits for everyone. The Pivot Kettle can improve people's kitchen experience by addressing water filling, lifting, carrying and pouring. It does this by offering a non-prescriptive handle design, providing posturally neutral hand positions to relieve physiological stress. Iterative development alongside physically impaired people provided a rich context for a relevant solution that appeals to a broad market. Furthermore, the design principles of the Pivot Kettle are ready for extension into other water products.

Samuel McClellan
Industrial Design and Technology

Sam Wong

An Inclusive Approach to the International Symbol of Access

Within a few decades the 'International Symbol of Access' (ISA) has become widespread across the world. Designed by Susanne Koefoed in 1969, it has been increasingly critiqued to perpetuate the stereotype of a disabled person as an adult male wheelchair user. Nevertheless the symbol has been successful in infiltrating the lives of those that come across it, yet we know remarkably little about this symbol and its purposes and functions. For those that recognise it, there are those who rely on it. Why is this symbol needed and what meanings affix to these needs?

The origin of the ISA stems back to the when the 'International Committee of Technical Aids' led the search for a valid symbol that would be simple, aesthetic, practical and identifiable from a reasonable distance. The aim was to establish an international symbol that would designate facilities made accessible to people using wheelchairs, that could be understood across the world.

There is a clear process in the conceptualization of the symbol from these facts. It was the president of 'Rehabilitation International' who sensed a communicative ineffectiveness of many differing symbols across varying foundations. This highlights the focused aim of the design process. The origin and the basic function of this symbol can be identified as a culmination of the needs at the time.

As with any graphical representation, especially commercial, the understanding, no matter how basic or crude its form, is a deciding factor in its success and so its interpretation. Hence its usefulness depends on the ability to recognise the symbol as a human figure on a wheelchair, and infer the symbol's relation to disability. There are many degrees of disabilities however and so many levels of need. How one symbol can represent a range of disabilities by only identifying one level is challenging to define.

The practical issue however of a single symbol to represent a group of people to eliminate confusion can be easily grasped. It is a logical process that fuels effective communication. Symbolically the ISA represents a specific disability but it is also simultaneously a metaphor for many other forms of disability. This metaphorical interpretation has evolved in physical environments but has expanded into online media. John Clark features an analysis into the Windows XP 'Accessibility Options.' He critiques the indication of online and offline accessibility and its relationship to physical space and mobility. "What does sitting in a wheelchair have to do with using a computer?"

The ISA provides us information necessary to understand an environment. In regards to navigating though a physical space and even cyberspace the function of the symbol has been adopted to provide a means to assist people by communicating the accessibility of a particular area. Moreover this accessibility lends to an individuals orientation within an environment, through use of nodes and signage as well as the construction of the environment itself.

This idea of the construction of the environment relating to a 'symbolisation of access' can be related to the concept of inclusive design. The natural progression of this topic leads then to the true meaning of inclusive design or the ultimate goal perhaps. These ultimate goals consider "the needs and abilities of the broadest range of potential users, and so reduces the necessity for such symbolic devices by reconstructing environments to have the fewest possible barriers. It is with no surprise then that the ISA provides a paradox through its own meaning and implementation. Announcing accessibility through the signifying of a disability denounces the lack in consideration to the environment it is situated in. Author of

'Design Meets Disability' Graham Pullin aptly writes, "People are disabled by the society they live in, not directly by their impairment."

Neil Cummings concludes that there is the potential to "use some of our commercial experience to breath some life into a tired and overlooked old icon," and the opportunity is there to change a one size fits all approach into a symbol that recognises people with or without mobility issues that they can feel proud of. The ISA however performs to so many roles within out society. Its success is highlighted through its wide usage, and its iconic nature has influenced individuals to modify the ISA in accord with the varying nature of their intentions. Many groups have adapted the symbol to emphasize the commonality of disabilities, and in turn their strength through the union in which they have created.

The modifications illustrate the strength of this symbol within the minds of those that it speaks to, as it performs to the basic requirement of a symbol that they utilise to represent themselves. To some, it performs to the extent of which pride can take form, but to others it brings with it a stigma so regularly prominent in our society.

The idea of disabled people sharing a single experience similar to each other is not the case, and likewise designers do not follow a single approach to design. It is the diversity in exploration that drives the creativity in the approach and development of new designs. An example is Lubna Chowdharry's beautiful ceramic tiles that contain specific colour palettes and textures. These tiles, blown up, configured within a grid give rise to the ability to outputBraille signage through changes in pattern and form. "These tiles could actually carry information in Braille," through touch as well as visually, and possibly inspire direction to approach signage. It is of no coincidence that

"People are disabled by the society they live in, not directly by their impairment." Graham Pullen

a process of thought and design has led to the potential of signage in a symbolic nature that can harmonise with its environmental surroundings, rather than paradoxically denouncing it such as that of the ISA. Another example is Stefan Sagmeister, a New York based graphic designer whom works closely with those he aims to speak to as well as working closely with people whose aims he admires. These are the basic principles of Inclusive Design, where the user is considered the real co-designer in the process.

Only through true collaborations between individual designers, development teams, and potential users, can a symbol create an experience that benefits the broadest range of people. It is with this idea in mind that the designs bought forth by those such as Lubna Chowdharry harbour the most engrossing creative possibilities. Graphical layouts have the potential to seemingly merge with environmental construction and in doing so possibly lend to the goal of inclusive design. Recognising our differences, through diversity, especially in design, it is important not to idealise whom we are, but more importantly understand our differences and how these differences can annex with each other. Inclusive design is defined through the inclusivity of its approach, not that of its ideals. Maybe this bears with it a truly creative approach and new ways to symbolise accessibility.

Seizure Protection Headwear
Protective headwear for people with Epilepsy

For people that suffer from severe and frequent forms of epileptic seizure, a medical protection helmet is an integral part of everyday life, helping to defend their head from regular impacts that their condition causes. Unfortunately, current design solutions offer little more than just protection. Poorly considered aesthetics and the use of cumbersome impact resistant materials often leave the wearer feeling socially isolated.

The challenge was to create a product which balanced protection and comfort with style and social acceptability. Use of the innovative impact resistant material D3O, allows for a flexible, breathable, thin-walled design solution. This headwear can now be a customisable medium, through which the wearer can express their own style and individuality, improving confidence and self-esteem.

Adam Smith
Industrial Design and Technology

Last year, 1 in every 400 newborn babies suffered from Cerebral Palsy. This neuromuscular condition can severely affect their ability to speak. Augmentative communication devices are used to aid communication, but many sufferers lack the motor ability and arm strength required to operate these devices. This project, run in collaboration with the Medical Engineering Resource Unit (MERU), focuses on producing a generic product that can be bought 'off the shelf' yet still alleviate the problem. The base unit and modular components allow the product to easily adapt to the widest possible range of users and environments without the need for specialist installation. Comfort and support is offered from the main body, but does not restrict the arm, leading to a more pleasant experience for both carer and user.

Rohin Odell
Product Design Engineering

Hayfever Aid

Reducing pollen inhalation and protecting severe sufferers

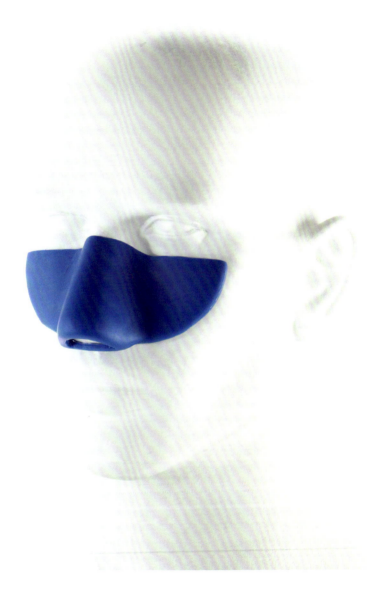

In a new, ever changing world, the need for solutions to improve everyday life is becoming more apparent each day. This project looked into how hayfever affects the quality of life of a sufferer and what methods can be used to improve this physical and psychological state. Hayfever is predicted to increase in severity and population percentage in the UK due to climate change. This being the case, there will be demand for new products that are more effective. This concept is aimed at 'severe' sufferers, and has been designed to act as a barrier, to be used both indoors and outdoors and for as long as the user finds necessary. The aim is to minimise the amount of pollen particles inhaled and so reduce the allergic reaction.

Ross Dexter

Industrial Design and Technology

Last year, 21,300 excess deaths among the UK's elderly population occurred during the winter months. This is the highest in Europe. Not only that, but for every death, there were eight admissions to hospital, 32 attendances at out-patient care and 30 social service call outs. The hard truth is that dementia sufferers are more likely to become one of these statistics. This is a condition that affects memory, thinking, language, judgement and behaviour. Warmfeelings is a temperature sensitive photo frame that provides alerts to the user if the room temperature is too cold. It utilises thermochromatic technology to display uninvasive messages. An added LED light alerts the user when the temperature drops below 11°C. Warmfeelings is inclusive and promotes independent living through informed decisions.

Olly Brown
Product Design

The Active Frame

Reducing loneliness in older people through ambient communication

Exercise Equipment For Seniors

A safe and exciting piece of equipment, designed for indoor use

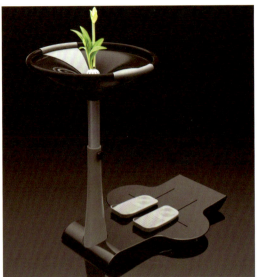

Lack of contact and interaction with friends and family can contribute to loneliness, depression, senility, Alzheimer's and dementia in older people. These conditions are expected to require an estimated extra £190 billion of expenditure each year until 2050 to manage healthcare requirements for the ageing population in the UK. The Active Frame is designed to reduce loneliness, using ambient communication with sensory functions, thereby enabling contact with friends and family wherever they are. Wealth within the older demographic is increasing and older people are becoming more discerning consumers and choosing high end brands. The Active Frame deliberately embraces a high-end brand, creating a truly inclusive product that anyone, of any age, would enjoy using, whilst being particularly helpful to older users. The Active Frame is a portable, multi-platform, remote-updatable, digital photo frame. It allows friends and family to send photographs from any location on a wide variety of platforms and also has the ability to sense when a person suffers a fall and offers the opportunity to identify their location. Further functions include a home direction pointer and automatic emergency messaging service. The Active Frame creates an inclusive solution mitigating the issues surrounding loneliness, often felt by older people, as a result of increasing distances between family members.

Exercise is an important part of people's lives and is particularly beneficial for older people. With the growth of the ageing population, the demand for exercise equipment for senior citizens will surely increase. The aim of this new product is to encourage people to do exercise regularly and make the process more interesting and safer for older people to use. The exercise equipment is based on the widely accepted concept of the 'air-walker'. An incremental improvement to this design is the introduction of multiple direction movement in both vertical and horizontal planes. As well as being adjustable in order to be usable by a wide variety of people, the design also combines a wheel exercising the upper body. A 3D projector projects a holographic virtual plant into the wheel which corresponds to the user's exercise. Energy spent on doing the exercise affects the plant like sunshine and water. In this way the plant will grow as the user exercises and will remain as a beautiful plant displayed in the home. Similar to gardening, frequent exercise will keep the plant healthy and continuously flourishing.

Nicholas Horan

Tingting Xiao
Integrated Product Design

There is a considerable gap of information in public knowledge about plastics. Since the introduction of recycling facilities, people have a greater understanding and can make connections about plastics in their everyday products; but this is still a tricky subject, which is confusing for many people. This project is an information platform to enable the public to learn and understand about the plastic products that they are buying and using, particularly when dealing with food. The most vulnerable people affected by plastics and the chemicals that can leach into food are child-bearing women, babies up to 3 months and the elderly. At any age, it is important to know what you are exposing to your body.

Jenny Schneider
Product Design

Future Concepts for Typhoo?

With over 100 years experience, Typhoo is one of the nation's leading tea brands and an established household name, offering their customers an escape from the stress faced in their daily lives. By aiming at the elderly population embracing travel, these new product concepts establish a closer link to nature.

These products, as a whole new package, are the bridge between human senses and relaxation of each sense: from the enhancing fragrance smell of nature to the sounds of nature, along with the comforting touch of a meditative Tai-Chi sword and stress relieving massage tool.

Pendant Scent Diffuser

Future concept for Typhoo

Life is a particularly long journey. With the sunset horizon of retirement, comes the sunrise of a new fresh open exploration waiting for you to discover. With a work free life, silver tsunami consumers have the opportunity to embark on a new discovery of unknown areas of the world. Accompanied by Typhoo, consumers are able to spiritually refresh and open their senses to capture the world's scenery. Typhoo has been able to keep its tea products fresh and revitalise consumers for many generations. Through the uses of fragrance and scent diffusion, this pendant diffuser enhances the consumer's experience of the world wherever they go. When desired, this device can be activated manually by the user to diffuse their chosen scent aiming to open their senses.

Nathan Chan

Industrial Design and Technology

"… an escape from the stresses of daily life."

The Typhoo brand focuses delivering relaxation to the customer and offering an escape from the stresses faced in daily life. Aimed at the older generation, when going on an exotic holiday or for a simple walk with nature, the user can carry this portable device. The sounds created in the environment are a major aspect intensifying the experience. When the device is placed open on a surface, highly sensitive integrated microphones record all ambient sounds. Once at home, the device is placed in the docking station, and replays sounds at superb quality, allowing the user to combine the pleasing sounds and remove unwanted elements such as automobiles or airplanes in the background. Now the sounds of nature can fill your home at a quality otherwise found only in professional environments.

Philip Zeitler
Industrial Design and Technology

Ty-Chi Sword
Future concept for Typhoo

Just as Typhoo tea aims to bring relaxation and the pleasure of fresh taste, other products can stimulate the rest of the senses in a natural way. Developed to stimulate the sense of touch, the Ty-Chi Sword is a dual-function tool that supports the traveller (and Tai Chi practicant) in their journey of discovery or spiritual pilgrimage around the world. Tai Chi is a Chinese system of slow, meditative, physical exercise designed for relaxation, balance and health. To correctly use the sword it must be seen as an extension of the body and mind, allowing energy to travel from the ground to the tip of the sword, extending the body's energetic field. The telescopic nature of the product gives the benefit of turning it into a foot massaging tool, created using the concept of reflexology, encouraging the traveller to self soothe when they need to.

Sorana Barbalata
Industrial Design and Technology

Our communities have become very disconnected from one another. People keep to themselves, and there is no longer an incentive to integrate with the local community. With constantly changing social nuances and attitudes, our daily lives are less about interacting with others and more egocentric. 'Encouraging Our Community' is a way of bringing people closer together. The initiative enables them to interact with each other and play a more active role in their local society. Royal Mail will introduce a range of services and products to encourage people in various kinds of communities to connect, interact and socialise with each other. This will increase well-being across communities, creating a close-knit and friendly atmosphere across the UK.

The NoticeBoard
Future concept for Royal Mail

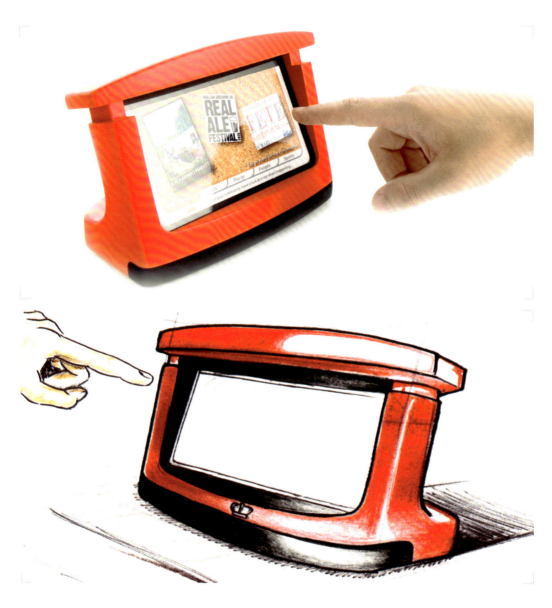

This project is designed to act as an interactive noticeboard for small rural communities. The device is primarily aimed at people who have recently relocated but do not get much time to socialise with their new local community or find out about events where they could meet others. The NoticeBoard will be located in various focal points of a village or small town, such as the town hall, local shop, post office, church and the pub. It will display information on local events, places of interest, people of interest, sports and so on, which will help the people in the community to gather information about what is going on more easily.

Jon Fletcher
Industrial Design and Technology

Apartment blocks are beginning to soar into city skies again in this age of incessant globalisation. Many people live packed into these buildings, rarely getting to know their neighbours. The communal garden is a space for occupants of the buildings to enjoy together, yet they are rarely used. As part of a new strategy, the Royal Mail will hold open days within the gardens of complexes encouraging residents to work together to plan and create their new communal garden. The CommunityBug is then placed within a plot and collates vital statistics on nourishment levels, streaming gardening tips straight to the user's laptop. The more residents that get involved or volunteer to kick-start the scheme in other complexes, the more points they collect, going towards materials for their garden's upkeep.

Steven Gutteridge
Industrial Design and Technology

Home Sweet Home-less

Future concept for Royal Mail

Due to high cases of homelessness throughout the country, London having approximately 3000 homeless people alone, it is obvious that in most cases people are being missed by the system. This project is about creating a stable, safe and empowering environment for people to live and get help to get back on their feet, to eventually rejoin society through work and counselling if needed. By social networking, it could provide people with a safe environment to work through troubled past-experiences, and give strength to individuals and groups as a whole. The proposed work scheme could provide job opportunities to Home Sweet Home-less participants and a small percentage is paid for the up-keep of their home. If this pilot scheme were to be successful in the UK, it could be expanded to other homeless people located around the world.

Jenny Schneider
Product Design

This concept aims to empower people with no home and no access to electricity. Allowing the owner to use, trade or sell the generated power it acts as a tool to create value and security. Devices such as mobile phones, shavers or radios can improve quality of life immensely and require very little power, yet access to energy is often denied. The idea of a cashless trade service completes the bigger picture of this concept. This device should embody a peaceful state of mind, the reassurance that it is possible to 'generate' the next meal.

Emanuel Köchert
Industrial Design and Technology

Community Calender

Future concept for Royal Mail

"Isolation can be a major source of distress for older people."

Isolation can be a major source of distress for older people. The Community Calendar is a service by which people can be connected to participate in mutually pleasing events. This may include road trips, outings, and lunch or tea parties. These interactive occasions provide opportunities for companionship and community feeling. Community Calendar works on a points system where the value of each event is displayed on a simple touch screen. Points may be accrued through online purchase, gifts from family and friends or purchased from the Post Office.

Joel Kemp

Industrial Design and Technology

A Poem in my Mail

A Piece of Creative Writing

The Mailbox is full, must get to work,
No slacking about, no figurative slur,
Happy the hour has come,
For me to deliver the love,
Lunch packed, keys in hand,
Pat the cat, out of the door,
Engine ignited, time to hit the road,
Flying to the beautiful shires,
With wind tucking under my hair,
Today I am delivering the mail

Knock, Knock, is anybody there?
Mrs B can't see well but she surely can hear,
Inside her living room, drinking tea and eating biscuits,
I ripped her letter open and began to narrate,
Her son sent her a postcard from borders afar,
Adventures he has had, such wonderment I grasped,
With a smile on her face, I took my leave,
To deliver more as time planned its escape

To his house and hers, I knocked and rang,
Big parcels and simple letters,
It didn't matter,
Their smiles made me well content,
He got a football, she got a guitar,
They got an invite, someone won an award,
She wasn't home so I left it in the letter box,
I delivered so many I lost count

As I passed through the towns,
In my fancy red van,
I saw people buying some stamps,
Putting more mails in the box,
What a nice way to stay connected, I thought to myself,
It might be outdated but it's still a comforting gest,
Such personal affection within them are held,
Comforting words and beautiful scents,
It will be a busy day tomorrow, I said to my pondering self,
More to deliver, and even more to collect

The bags went empty, parcels all gone,
A job well done, must head to home now,
Put my feet up, contemplate a little,
Satisfied customers, a hot cocoa is well deserved,
A voice from the forgotten past echoed in vain,
Silence befell on my lonely domain,
Another day went by, no letter for me,
When will she let me know that she misses me?
Days into months, months into years,
Time has flown by beyond recognition,
But not a worry, I will wait for it a bit more,
Till then I shall keep on delivering the love.

By Momus Najmi © 2011
Creative and Professional Writing MA
momus.najmi@gmail.com
07412 282044

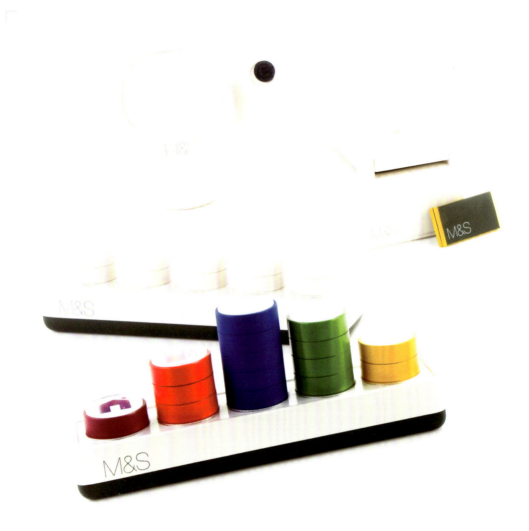

Since 1894 M&S has been built around the five key principles of quality, value, service, innovation and trust. It has grown to be one of the UK's most trusted high street brands. They already financially support charitable and community based programs, however their trust could also be used as a foundation for bringing people in communities closer together. The term "community" can be interpreted in many different ways. The strength of communities can be directly increased through M&S supporting neighbourly and personal relationships, family matters, skill sharing and financial management. This range of future concepts for M&S addresses a range of issues, but all have the same aim of strengthening the trust that holds relationships together.

Connect

Future concept for M&S

This small device is a replacement for the humble front door key of the family home. The device recognises each family member's finger print and opens the door as they approach it. The device's secondary function is to make communication easier between family members when they are away from the home. When family members are apart and they would like some reassurance that others are well, they can squeeze their device, which will cause the others to glows or vibrate. When the device glows, the user can 'reply' by squeezing their device. This simple action gives reassurance for all the family members without invading their private time. It is quick, simple and effective.

Katy Koren
Product Design

Groceries, gas, water, electricity, broadband, transport, rent or mortgage. We all have personally important living costs. We need these services, and we have to pay for them. It is hard work managing them and they can build up to worrying figures if ignored for too long. The current ways we manage our finances are through mail, phone calls, emails and websites. If these are put to one side, negative coping strategies can create stress and extra mental effort. Physical Finance improves money management skills by providing a physical interface that encourages positive actions. Each cylinder can be assigned to a financially important aspect of your life, indicating relative size, timing and importance. Physical Finance is private when it needs to be, yet informative and positive when the right person is nearby.

Nick Sardar
Product Design

Share
Future concept for M&S

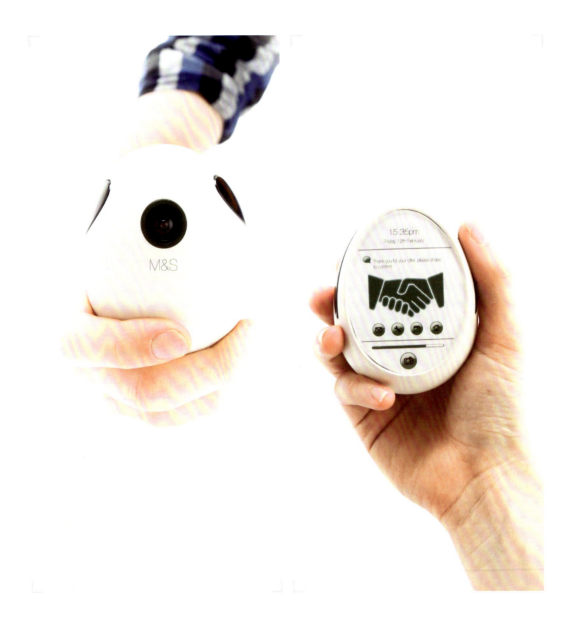

'Share' is a community based bartering system aimed at facilitating communication between people in the local community. It provides the user with a platform upon which they can borrow, swap or share anything they choose, from unwanted items to skills, trades and talents. By providing the consumer with a non-monetary currency to trade, the device aims to re-instate the lost sense of community within society, by encouraging people to work with and for each other for mutual benefit.

Adam Smith
Industrial Design and Technology

There are many people out there who, for many reasons, prefer to shy away from direct communication with others and keep to themselves. Some find this frustrating as it is not how they want to be, but they have difficulty commmunicating their true personality. M&S Pen Pal is a means of indirect communication that encompasses a personal feel. A transparent smart-postcard can be used to capture an image, superimposing it onto its surface and becoming opaque. The user can then write a message on the back of this and post into the accompanying post box. The message is absorbed, sent into "the cloud" and an appropriate receipient is found based on the information on the card. This can act as a means of starting a meaningful connection between individuals and, in the future, could lead to them meeting face to face.

Max Woźniak
Product Design

Neighby

Future concept for M&S

In many areas within the UK, there is a lack of community spirit. Neighby encourages good deeds within the community by providing a credit reward system. Each community member will have a Neighby in their household. If they are rewarded a credit, their Neighby will produce a coin which he will hand to his owner. Credit coins can be used as Marks and Spencer vouchers, giving community members the incentive to show good will to others within their community.

Dee Fisher
Product Design Engineering

In response to older people feeling isolated more than ever from each other, the M&S message box provides an enjoyable means to send and receive stories and experiences to help people understand our differences and celebrate diversity in age. The box wirelessly connects to an online space and provides physical representations and notifications of messages sent and received online. When dormant, the box is just a box, but when active it blossoms like a flower to represent growth and wisdom and reveals hidden messages that have been sent in reply to journeys and curiosities we experience throughout our lives. Through understanding each other, and our differences and similarities at whatever age, we can build on the foundations of younger and older generations growing up together, not apart.

Samuel Wong
Product Design Engineering

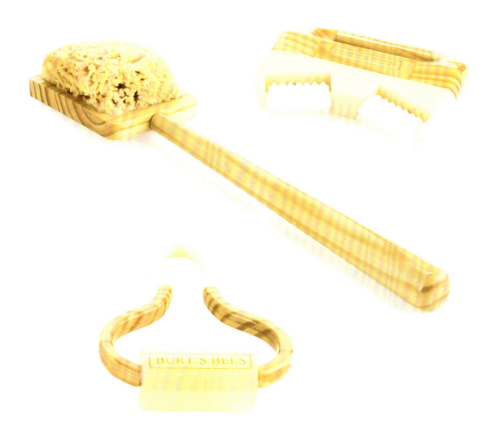

Burt's Bees is an ecologically conscious brand specialising in natural personal care products. They will continue to be green by anticipating future needs of their customers and understanding the environment. Working within Burt's Bees' values and missions, these concepts will develop and expand the company's men's healthcare range. This project investigated potential trends in society as well as the design industry. Environmental concerns are key to our thinking and new products should aim to be being sustainable and natural. These products are designed to raise the natural quality of life by improving the human lifestyle. The future trend for green products is going to be led by Burt's Bees to provide better living for everyone.

Shower Scrub

Future concept for Burt's Bees

Burt's Bees are experts in the field of natural personal care products. In-shower products are daily necessities for everyone. A shower scrub made with 100% natural materials will be a potential entrant to the male grooming market, as it improves personal hygiene and helps to build confidence. Burt's Bees' products are natural, sustainable, unique, honest and inclusive. Natural materials correspond to a tough image, which will be a feature that increase the appeal as a male grooming product. The materials used are environmentally friendly. The inclusive design provides features that are easy to use and which fit the needs of a wide range of men. The style, features and aims of the product respond to Burt's Bees' mission of making only natural products and encouraging our sense of wellbeing.

Michelle Poon
Product Design

This wash bag was designed for Burt's Bees to target male consumers for travel or sports activities. The design follows the brand essence of Burt's Bees; "Natural, Rough and Simple". Therefore oak, linen and hemp are the main materials used in this product. The lightweight bag is easy to carry, and can be rolled up and secured using a wooden toggle. Each pocket of the bag is triangular to hold the bottles, making it easy to store. Twist and press the bottle cap to open; this is an efficient way to ensure it does not spill.

Bing Wang
Industrial Design and Technology

Natural Massager

Future concept for Burt's Bees

Today many people suffer great pressure from a fast paced working life. Our well-being reflects not only on physical fitness but also on the mental and social aspects of an individual's life. Relaxation is important as it helps to keep stress levels down and consequently improves our health. High-tech relaxing devices on the market function well but lack emotional value for the user. Burt's Bees can marry their natural personal care knowledge and environmentally friendly values to develop masculine relaxation products. The classical hand tooling keeps the masculine appeal while healing the heavy fatigue from work. Let Burt's Bees soothe your body and revitalise both life and spirit.

Zhen Lun
Product Design

This is a soap roller for male grooming. Wake up in the morning with an invigorating buzz. The Burt's Bees soap roller is recommended for everyday use for the face and body. It was designed for a man's hand size and with a handle made of ABS plastic provides an effective solution to the age-old problem of soap slipping. It has a maximum flexible range of 5cm. The soap in replaceable and once the soap runs out, simply change to a new roll. By reusing the handle each time, the product is low-carbon and environmentally friendly.

Meihui Yan
Industrial Design and Technology

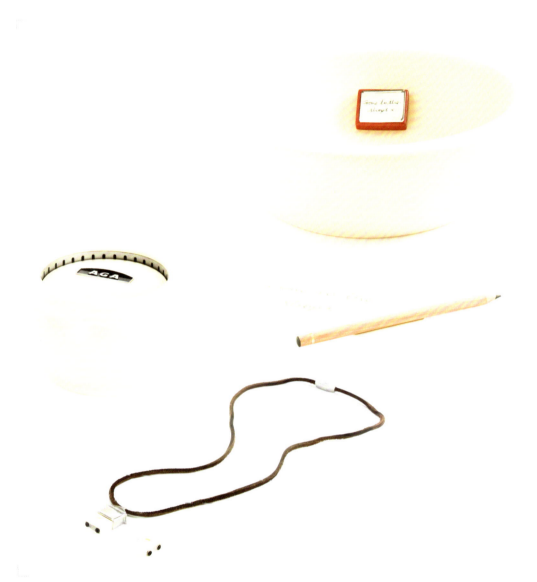

The combination of fragmented families and hollow relationships create an opportunity for AGA to apply their unique ability to bring people together in a modern context. The physical warmth an AGA has always provided to families can be applied in a new way. The warmth can instead also be provided emotionally, by improving the quality of social interaction.

The product concepts put forward are not to replace the AGA cooker, but coexist with it. They will continue to carry the core brand values into 2025, keeping them relevant for people whose lifestyles mean they are unable to spend much time in the environment that the current AGA range suits so well.

AGA Watering Pot

Future concept for AGA

The AGA Watering Pot aims to help the growing problem of a widening generation gap within families; often leading to elderly people living alone with the risk of feeling isolated. Feeling needed is one way to help this, as having something to nurture becomes important after the children have left home. The concept is a thumb watering pot for separated elderly and younger family members to own. Watering and caring for plants is a daily routine; the bowl includes a sensor to record how often the pot is used. This information is communicated to the other family's bowl, giving them a reassuring indicator that their loved ones are okay, and also a subtle reminder to get in touch. The design of the watering pot is based on a forgotten 16th century design to embrace AGA's nostalgic and utilitarian qualities.

Mike Puttock

Industrial Design and Technology

The AGA Pendant is not just a family heirloom, but like an AGA, it is a power source to give energy to the other areas of your life. The master pendant powers the various sub pendants. These sub pendants are valuable for times in life when the family physically cannot be together. The house pendant is to given to children when they first leave home, and will glow with a gentle warmth to show they are being thought of, reminding them that they are welcome home anytime. The second egg pendant is symbolic of new life, it will pulse to the same beat as your newborn's heart. Whether Dad is at work or Mum is in bed, by triggering the sub pendant they can physically feel the heart beat of their newborn. A sub pendant is attached to the main pendant by a magnet and draws its power from the master Pendant.

Matt Rowinski
Industrial Design and Technology

AGA Scent Pot

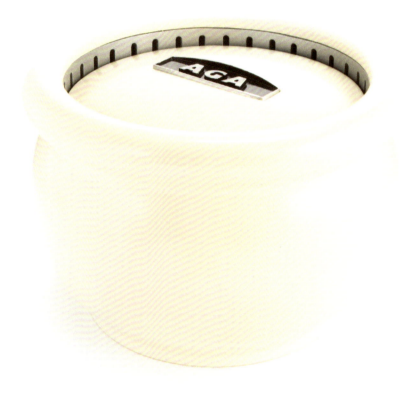

Moving out of the family home is a daunting experience, it is incredible the little things you miss. The AGA Scent Pot reproduces the traditional aromas of home based on the time of the day, from freshly made coffee to homemade bread. This helps to evoke a sense of emotional warmth and familiarity within the home. The pot brings forward the classic elements of the AGA brand such as the vitreous enamel finish, but subtly integrates modern technology to make it an attractive product for future consumers.

James Willson
Industrial Design and Technology

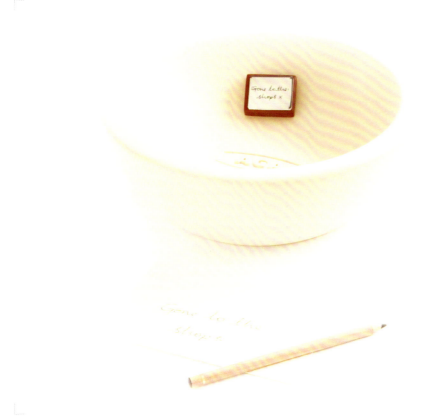

The product concept addresses the issue of decreasing communication and the quality of social interaction within a family. The concept aims to do this in a subtle, unforced way, to help family members regain the emotional warmth achieved by spending quality time together. The product uses technology to pass information between family members with the hope that this will encourage them to communicate when they next see each other in person. The main hub of the idea focuses around a bowl, placed somewhere central to the home, that all the family can use to store everyday objects such as keys. Each member of the family has an AGA key ring. When someone leaves a note at home, or an important event is near, this is then sent electronically to the relevant family member's key ring, and everyone is kept in touch.

Jonathan Rugg
Industrial Design and Technology

Shoulder Support
Preventing shoulder injury within whitewater kayaking

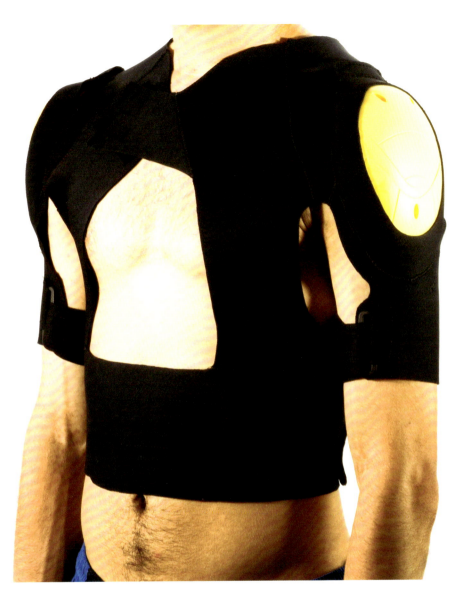

This protective shoulder support has been developed primarily for use within whitewater kayaking, although it also has potential within other sports. The shoulder has the greatest range of motion of any joint in the body, however this results in an inherently unstable joint. Sports often require this full range of motion for optimum performance and this can frequently lead to injury. The two most common causes of acute shoulder injuries are wrenching the shoulder joint due to an external force, and trauma due to a direct impact. This protective shoulder support is designed to prevent shoulder injuries resulting from both of these causes, whilst ensuring the freedom of movement is not significantly reduced. This is achieved through a bracing neoprene support with integrated shoulder protectors to absorb multiple impacts.

Matt Smith
Product Design

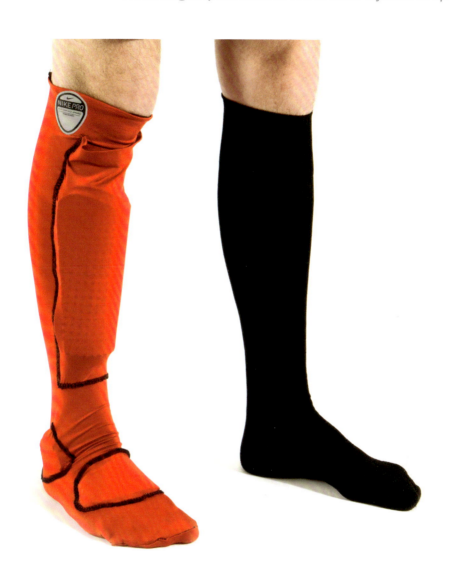

The occurence of lower limb injuries due to impact in sport is on the rise. This is due to a lack of protection in the sporting footwear currently on the market. SIRIS is a Sports Injury Reducing Impact Sock. The product is designed to incorporate new impact materials, yet still retain its form and be comfortable for the user. SIRIS provides a new direction in sporting protective equipment offering a much needed solution to a problem that is rapidly increasing. SIRIS utilises impact material 'D3O' alongside the stretch sports fabric 'Dri-Fit' which is incorporated in many of Nike's sports clothing products generating a lighter and more protective product.

Jamie Trigg
Industrial Design and Technology

Athlete Endurance

An ankle weight for track and field athletes

This product is a development on existing ankle weight products specifically designed for track and field athletes. Every aspect of the product has been carefully thought out to make it as effective and comfortable to use during long and intense training sessions. With more consideration to ergonomics and human anthropometrics rather than the 'one size fits most' approach commonly seen in the design of existing products. This product provides its users with a snug fit stopping any repositioning from occurring during training. The use of Spacer fabric for the construction of the product provides added durability as well as breathability, greatly reducing the amount of discomfort caused by a build up of sweat between fabric and the skin.

Yinka Branco - Rhodes
Product Design

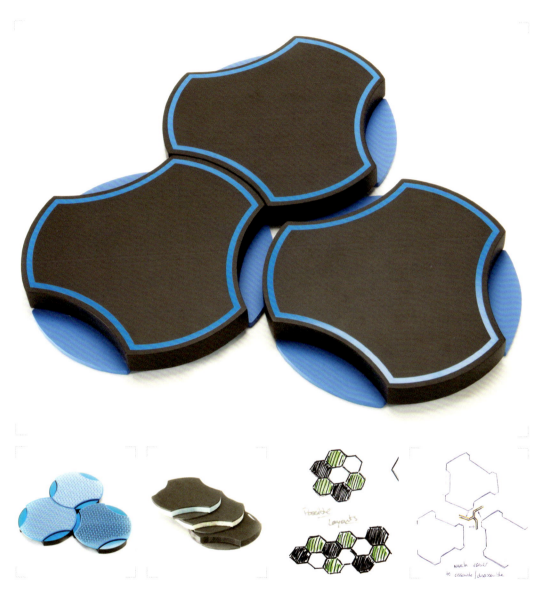

Ankle sprains are a significant problem in society, and are most common in athletes, children and the elderly. Due to decreased proprioceptive function post-injury, ankle re-injury is common. Proprioceptive training has been proven to be beneficial both in regaining lost proprioceptive function, but also in its development before initial injury. This concept is based on the idea that an unstable surface underfoot can stimulate the necessary sensations for proprioceptive development. Using closed-cell foam, slight instability can be created without increasing the risk of injury to the user. The modular mat can be arranged in any shape or size providing random instability suited to the needs of each user. Design focus is on portability, weight and ease of implementation into existing lifestyles of both healthy and injured users.

Alex Williamson
Product Design Engineering

Pro Ring Bottle
Controlling a healthy protein intake

Many protein shake consumers are damaging their health by consuming too much protein. Most users are unaware of the fact that there is a maximum amount of protein that they should be consuming in relation to their weight. The Pro Ring Bottle is a product that is designed to solve this issue. The bottle incorporates a rotary calculator, allowing the user to select their weight and discover their maximum daily protein intake recommendation. The bottle also includes a motorized blending mechanism to solve the common powder clumping issue that comes with most protein shake bottles.

Dee Fisher
Product Design Engineering

Delayed Onset Muscle Soreness (DOMS) is the pain experienced after undertaking heavy exercise. It will usually appear 24 hours after completion of the exercise and affects sports men and women of all abilities. The most common method of muscle recovery by athletes and physiotherapists is the ice bath. Research suggests that people find the experience highly unpleasant and time consuming and it is not a realistic option for the sports amateur. Compress is a piece of sporting apparel that will treat lower body muscle soreness using custom-made 'icing gel' inserts. The inserts consist of welded divisions that separate the gels, creating a more flexible pack that can fully compress the user's body. It can be used just as effectively to treat injuries and can also be used as performance underwear during sport.

Matt Parrish

Product Design

Lotus is known for its range of sport cars designed for the person seeking the ultimate driving experience. Lotus has expressed a growing interest in greener technology, with the release of products engineered to enhance the thrill of a journey, but using as little fuel as possible to achieve it. Lotus is not the brand for gimmicks, they are a brand relying solely on the excellence of their engineering and most importantly the experience their products give to their customers. Extreme Self-Propelled Sports is an experience in itself, relying on a natural power source: human strength. The tools required for such sports need to be well engineered for function, but also designed to heighten the thrill, excitement and intensity of the extreme sporting experience.

Lotus Mountain Climbing Helmet

Future concept for Lotus

The brand Lotus represents the epitome of engineering excellence providing the ultimate experience for its clients. With green being the new black and Lotus showing an interest in greener technology it would be only fitting for Lotus to produce a range of equipment for extreme self propelled sports. The product chosen was a mountain climbing helmet. The unique shape of the helmet is inspired by current Lotus vehicles. The material used it called Dyneema HB80, the world's strongest fibre, an ultra strong polyethylene fibre that offers maximum strength combined with low weight.

Seun Babatola

Industrial Design and Technology

Lotus is about the experience, the thrill and beautiful engineering that works. A good ice axe is lightweight but heavy enough to control, strong enough to withstand the impact of penetrating the ice mountains and engineered for the climber to get maximum anchor into the mountain caps. The Lotus Ice Axe is the pillar of the intense Lotus experience. With green being the new black, the Lotus Ice Axe aids the self-propelled extreme sport of Ice climbing. The shaft is Titanium, high in strength but light in weight, with the handle of the ice axe made of a thermoplastic rubber for maximum grip.

Efua Mensah-Ansong
Industrial Design and Technology

Lotus Diving
Future concept for Lotus

Lotus Extreme Sports presents Lotus Diving, the first wetsuit to monitor the swimmer's performance. The training suit is designed to record the stroke rate, heart rate and with a built in stopwatch to record lap time. The wetsuit uses a heart rate monitor incorporated into the suit. The signal travels down the arm to the monitor along the forearm for the swimmer to view their figures. The training suit can be taken into more extreme environments with the implementation of a unique smart material called D3o which allows the suit to flexibly glide with the body, however, on impact the suit changes its state, locking its molecules to absorb the energy shock. Lotus Diving is designed to enhance your swimming performance through its advanced engineered suit as well as give you the extreme water experience.

Hannah Haile
Industrial Design and Technology

In the future, Lotus may develop a range of extreme sports products. These sports would be self propelled, non-motor based, sports in response to the current concerns for the environment and the damage from vehicle emissions. These sports will allow the user to achieve that Lotus rush and the charged experience that they would get from the car, but in a different sporting medium. The Lotus Climbing Belay provides that sustainable excitement through climbing, whilst conveying engineering reliability and quality.

Mark Connor
Industrial Design and Technology

Chris Strickland

In the creativity game

In the modern world new technologies are everywhere. Constant changes through competitive development drive down prices, and technologies encroach on our daily lives. Try as you like to avoid the iPhone, Facebook or the latest home electrical, somehow the unstoppable march of development will get you in the end. Companies can no longer rely on competence alone to ensure they bring competitive products to market, it also requires creativity.

Sherlock Holmes famously remarked: "When you have eliminated the impossible, whatever remains, however improbable, must be the truth". To rule out the impossible, the realms of the possible must first be explored, this is where techniques such as brainstorming and 'blue sky thinking' come into play. Other internal factors have an important, sometimes unexpected, effect on the ability of people to be creative. These factors can range from a person's psychology, personality traits (risk taking, perseverance etc), motivation, and creative thinking abilities.

Teresa Amabile of the Harvard Business School argues that these internal, intrinsic, factors encourage creativity and external forces restrict it "Within every individual, creativity is a function of three components" These are: Expertise - the domain specific knowledge that must be acquired in order to be aware of the existing work in a field; creative thinking skills - psychological methods that make a person creative; and motivation; which may be internally or externally driven. Motivation can be the most important factor for creative thinking.

Creative skills essential for success can be encouraged through the use of lateral thinking techniques. If everyone can develop their creativity skills, everyone should be successful in any creative endeavour they apply their skills to. This may be true, but the skill must be coupled with the relevant knowledge which allows them to 'think outside the box', for to think of something new you must know what exists already. Knowledge, particularly domain specific experience is essential for creative ideas to be widely successful. Although having knowledge of past successes and failures may allow you to use them to your advantage, the mind may get stuck in an existing pattern of thought, and so be closed to all other possibilities.

To truly get the maximum from creative people they must be correctly motivated. Teresa Amabile has identified that motivation comes in two distinct forms: intrinsic and extrinsic. Extrinsic motivators are the external factors, (such as monetary reward, time pressure, or punishments). Intrinsic motivation comes from within, a personal interest in the project. Amabile presents intrinsic motivation as most effective for creativity and extrinsic motivation as detrimental to creativity.

Practical support for the creative process in business is important at every stage. Through extensive studies of creativity in business, Amabile identified key management methods that help foster creativity through intrinsic motivation. One of the simplest things to implement is 'matching challenges'. Getting this right requires managers to understand the personalities and motivations of each of their workforce. Effectively actioning this means the individuals are personally motivated to complete the task in hand. It is however; the most common way managers kill creativity, simply ignoring the individual abilities within their team.

Freedom comes in many forms; the freedom for anyone to suggest ideas in a meeting, the freedom to work when best suited. Some companies excel in this regard. One example is the renowned design and creativity agency IDEO. General

Manager, Tom Kelly, encourages his design offices to have their own identity, and each designer to personalise their desk space. Kelly's philosophy is the fewer rules the better, an office space should reflect the people in it and the people in this office like to do things differently, enforcing the organisation's values of openness and, importantly, creative thought.

Resources, both time and money, often prove a sticky issue in any environment, but when badly managed in a creative environment they can be especially damaging. Unfortunately, they also tend to be the two things that are in short supply. Amabile identified that in most situations time pressure has a negative effect of creativity. A prime example of the exception to the rule is the familiar story of Apollo 13 when NASA engineers had to improvise a method to fix the air filtration system. Under this kind of pressure people feel a sense of purpose and are motivated to complete the mission; "A successful failure", that led to some ad-hoc invention. Maybe the only reason that IDEO, and others get away with providing complete flexibility is because they have succeeded in motivating their staff completely intrinsically, so their staff believe in the work that they do.

Does competition beat collaboration? No. Although competition may create many ideas, the creative process must embrace collaboration. The success of creative work is affected by the amount and type of collaboration engaged in during the day: "Evidence adds to the picture of a distracted, disturbed, confusing environment on treadmill days. People had more meetings and discussions with groups rather than with individuals" says Amabile. Back at IDEO, Kelly encourages and witnesses off-beat collaboration all the time, as they regularly engage in unrelated team competitions. Kelly suggests that this kind of group competition can help build team cohesion, an important aspect of the team dynamic.

"Fail often to succeed sooner" - by embracing failure in IDEO's process Kelly is able encourage the whole company to be creative. If there is any one person at the table that is not prepared to change, then the whole company could get stuck. If CEOs and executives frown on failure, a stigma can develop; people will no longer be willing to take risks.

Praise and intrinsic reward are the foundations for success. Continued encouragement during a project and recognition of all successes, before the end of the project is important for the building of successful teams. Managers regularly kill creativity by focusing on the negative with critical reactions. This leads to an environment of criticism and the focus transferring to extrinsic reward.

If businesses are to push the boundaries and keep the progress of product development at its current rate, then they must embrace systems to foster creativity. Companies must build creativity into the fabric of their corporate structure. Taking well-considered risks is essential for progress. Embracing the three core factors influencing creativity, will lead to companies being better able to compete.

The danger however is providing the wrong kind of motivation. If a system built on extrinsic motivation, such as bonuses, is relied upon the benefits of all other attempts to foster creativity may be lost. Far better to spend the money providing team activities to build team cohesion. To improve their creative output companies will need to effectively manage and develop the knowledge, skills and motivation of the designers who create products. Companies can then take bigger innovative leaps, and force us all to run to keep up. As Sir George Cox puts it: "The outcome won't affect just business itself. It will affect us all."

Converse have never followed trends. Their story is one of new beginnings and ideas. When the original shoe appeared it did not follow convention or look and feel like anything else; it was just Converse. And now, over 100 years later it is still as fresh as the day it was first made. With each new generation, a new wave of Converse people usually make themselves known. We saw it with basketball, we saw it with punk. It is those who stand up and say "I want to try something different." The Converse of the future will help people create, share and express the moments and ideas that make them who they are. The highs of a defiant lifestyle will be transformed into tangible entities which can be shared, collected and combined into dynamic displays of youthful rebellion.

Rec Deck
Future concept for Converse

This carbon fibre skateboard boasts six high speed cameras mounted inside the trucks to capture skate footage from an entirely new perspective. Since skateboarding tricks are all about the movement and rotation of the board, these cameras will capture every flip and spin in a way that has never been seen before. Once filming is complete, the board is flipped over to reveal a touch screen interface on the underside of the deck. Here, stunning videos are created by editing together footage from the different camera angles, adding with super slow motion replays, dropping in soundtracks and so on. Rec deck brings people together in a very real way, uniting their dynamic, creative personalities and having loads of fun at the same time. Videos can then be saved or shared as a lasting testament to the Converse revolution.

Benjamin Whitehead
Product Design

Converse has always been revolutionary, right from day one when it differentiated itself from its competitors. There is a rebellious streak that runs through the brand and empowers individuals to express themselves for who they are. Focusing specifically on graffiti, an urban art form particularly apt for Converse, this conceptual product uses Augmented Reality and Virtual Retinal Display (VRD) technology to create virtual graffiti. The two devices, a hand-held Converse spray-can and VRD head-set, would allow the user to view and create pieces of artwork and leave their feedback. This is all done within the 'Converse Community' so the graffiti would only be viewable to individuals with a Converse head-set. This allows like-minded people to form a point of view, share their opinions and forge new friendships.

Emma Tuttlebury
Product Design

Detonator

Future concept for Converse

Ubiquitous computing is coming. Technology will be everywhere, and everything we own will be connected. We will have no need for physical interaction, resulting in frequent seclusion. This will have a significant impact on the nation's social well-being. Converse is beginning a Revolution. It is time for people to stop spending all their time in front of computers. Whatever people enjoy doing, they should go out and do it together. Converse has always been a rebellious brand and the Detonator epitomises this. One press sends out a simulated EMP blast that shuts off all the technology in the room and sends out an alert to the detonators of friends. A laser touch interface allows the users to arrange a place to meet and can direct them there. The Detonator encourages social change and brings the Converse community together.

Adrian Hodder

Product Design

Community and music are two important characteristics of 'Converse Culture' and the drumsticks target these by bringing people together through urban drumming. Urban or street drumming involves percussive drumming on anything and everything in the surrounding environment. Switch on the sticks and they automatically sync with friends and others in close proximity. Then the lights on the sticks give visual cues and a performance can be achieved with no prior musical training. Each pair of drumsticks gives an individual beat to follow, enabling a group of any number to perform. The loud, percussive sound can be created from anything; bins, railings, benches, bottles... The drumsticks promote social interaction within an urban environment through the rebellious and physical action of drumming.

Matt Smith
Product Design

Converse Jax

Future concept for Converse

Jax is a conceptual product for Converse, with the principal aim of improving people's physical well-being as part of a socially engaging outdoor environment. Aimed at the younger generation of the 2025 Converse user, the Jax game adds a new and challenging dimension to the urban sport of skateboarding, providing an exciting and creative means of exploring your skateboarding talents. The set of six ruggedised Jax pieces each emit a beam of light that once broken by movement will register as a captured target. The Jax can be grabbed, twisted and thrown to make creative target pathways; setting challenges between friends, and encouraging imaginative new locations for the Jax. Remote connection to the portable feedback interface allows review and comparison of target capture performance inspiring healthy competition.

Jade Boggia
Product Design Engineering

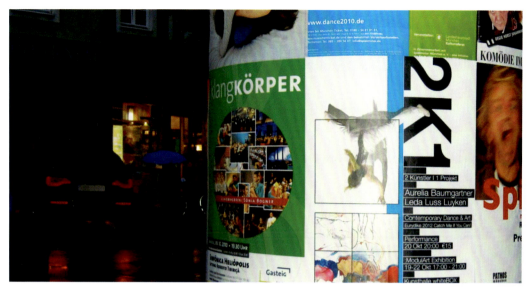

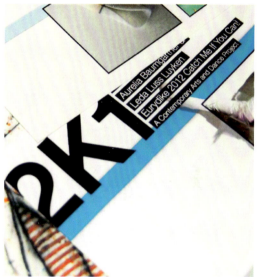

LLLarts is a company that features contemporary conceptual art with a clean-cut structured approach. The newly created brand focuses on an artist's professionalism in the creation of innovative conceptual projects. This incorporates videos, fair trade booths, multiple artistic events and performances, posters, as well as a complete stationary set consistently into a both appetizing and recognizable format. This format rests on a strong visual design language that is artistic in style, contemporary in expression, and easy to understand in its delivery.

Roman Luyken

Product Design Engineering

Constructables

Turning children's play equipment into an infinite play environment

It cannot be denied that society is changing rapidly, and with this there is the danger that we lose some of our more traditional skills and values, and this is of particular importance for the next generation. 'Constructables' aims to free children from 21st century constraints and encourage them to play in a more traditional manner. While it primarily serves as child's furniture, it is intended to be disassembled and reassembled by the child as well. By allowing children to configure parts from a stool and table in many combinations to create structures, it aims to encourage child development through the play associated with den building and construction toys. Constructables is a fun product that will help to create well-rounded and happy individuals equipped with the necessary skills to face adulthood.

Lucy Alder
Product Design

This is a collaborative project with Peter Black Footwear & Accessories (PBFA) aimed to produce a set of infant's toys that are made for children's learning strategies. Parents' spending on toys for their children is generally increasing. The reason for this is that children are losing interest in toys quickly and many only cover a short period of their development. The aim was to produce an effective design to enhance infants' senses and support different areas of children's development through their rapid growth. The target users and customers are infants aged from 12-36 months and their parents. The toy will hold the infant's interest for longer through multifunctionality which may progress at different stages of development.

Michelle Poon
Product Design

Autoy

Social skills toy for children with autism

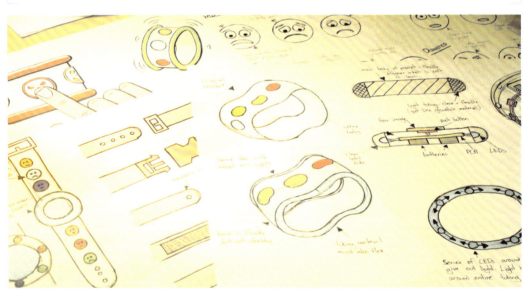

Autoy is an educational toy which encourages social skills in children with autism and Asperger Syndrome. These developmental disorders affect one in every hundred children in the UK, yet very few products cater for their specific needs. Using Autoy, the child can select their current emotion by pressing the corresponding one of six faces around the wristband, causing the rim to light up in a linked colour to communicate this emotion to others. The product helps children with autism express their own feelings, and recognise emotions in other people too, which makes them more socially adept. Child to child interaction is increased as well as child to carer interaction so parents and teachers will better understand their children. Autoy also comes with Feeling Tracker software which allows parents to review their child's recent feelings.

Benjamin Whitehead

Product Design

Research suggests that hi-tech toys do not supply educational benefits to infants, yet the current toy market is led by PC and console based electronic games. This project aims to redesign and manufacture a traditional toy, targeted for infants. The project involves understanding current toy market issues and the potential benefits of a Tangram. It also involves individual consideration of the original product's specification. The main task is to increase the popularity of Tangram and approach the consumers' needs. This project involves new more creative ways of playing the Tangram with a redesigned structure.

Bing Wang
Industrial Design and Technology

Interactive Infant Play Mat

Encouraging and enticing babies to move and crawl through play

Children learn through play which is important for their developmental growth. Babies learn to crawl at different rates during their development. This concept encourages the baby to move around more actively, from initially rolling around to crawling. With the use of embedded switches, when the baby moves around on the play mat this will trigger another area of the mat with an interactive response, encouraging the baby to respond to this and want to move toward it. These interactive responses are actuated by small motors that produce soft sounds from rattles and beads. The mat also has surface features that will attract and encourage the child to approach these sections, including different textures and shapes. This is so that the baby has many elements on the mat to give them a most enjoyable play experience.

Turgay Hassan

Product Design Engineering

The Hornby Memory Maker brings a new twist to nostalgia, allowing families to rediscover their past like never before. Simply select the elements of a memory such as the people, places, or events as represented by coins. Insert these into the tray, slide it into the Memory Maker and let the device select the memory that is most applicable.

Timothy Dunkley
Product Design

Hornby

A Piece of Creative Writing

The mind's a funny old thing ain't it? Get to my age and you have to second guess yourself all the time. I wouldn't say I've got Alzheimer's, but I'm definitely a bit rusty. 'Senior moments' is what my Tara called them, 'Being a daft old sod' is what I told her. She'd watch me turn the oven on to make a cuppa then burst out laughing. We were like that, me and Tar'.

Now she's not here and I'm pushing eighty it ain't so funny. I went down O'Dowds on Christmas Eve and the place ain't even there! It's a carwash now, such a shame. Great pub it was, and all them characters! Comical it was; I wonder what happened to the framed photos down there? I wouldn't mind taking a look over them; try and jog a few memories out.

I have to wait for a memory to find me these days; a certain smell or something. West Side Story was on the box a while ago, took me right back it did. David on his wooden stool, helping Tar' wash up singing 'I wanna be an American!' in his apron, Tara looking back and smiling that smile. I wish I had a photo of it.

The film was on the telly again Christmas morning when I went over to David's. I was thinking of that scene when I watched him open presents with his own sons, Joshua and Harry. Truth be told we don't see eye to eye, me and David; he's never got the time really, what with the job and the kids. She's a great girl his Sandra, and good on her having two kids in her forties. Bringing up one nearly done me in. But, yeah we're different. He's a clever sod for starters, but sometimes life's not about exams is it? He talks to me like I'm a bloody idiot, even Sandra speaks like I'm not even there sometimes.

They thought I didn't know they'd been arguing before they picked me up, but I could tell a mile off . The kids who were normally screaming the house down were sat back watching the film. Sad sight on Christmas day.

They brightened up a bit when we sat round the tree, but then Harry went for the biggest present… Well, I don't remember ever telling David off like that!

'Wait your turn Harry!' David yelled. 'Don't you ever listen?'
'Don't talk to him like that!' Sandra barked.

On they went, what a racket. While they were at each other's throats, the cheeky sod opened it anyway! Turns out it was from me. Well, I'd sent Sandra a few bob in the post. Smashing present it was, a bit like the train-set I had as a boy, but with them smiley faces like on the telly. Reminded me of my dad and uncle shovelling snow on the roof one Christmas when the ceiling started to bow. In they come, sodden wet, got down to their smalls and sat by the fire. What a sight: two silly sods in their pants, smiling away at me playing with my new train set.

After watching Josh and Harry push their dinner round their plates, sulking as much as their parents, I'd had enough. I went and got my grandson's gift, creaking and cracking as I went, and managed to sit back down at the table and put together a bit of track.

'I don't think that's appropriate at the table, dad,' David said.
'Knockers,' I said back.
Even Sandra laughed at that.

She joined in, laying the track past the gravy boat to Joshua and then back round to Harry who was giggling like a good 'un. David put the fancy candle-bridge, the menorah or something, over the track, and when the train chugged under the kids both cheered. Tara would have loved it.

By Johnny Edwards © 2011
Creative and Professional Writing MA
johnsemailadd@aol.com
johnnyedwardswriting.wordpress.com
07908763065

It would be hard to find a board game in a family home that is not produced by Hasbro. Their leading expertise means that they have developed a dominant position in the pre-school product and board game market. It is clear that Hasbro believes in the notion of 'fun' and 'laughter' and with this firmly imprinted, the brand direction developed sees Hasbro produce its own range of toys to help children develop.

Developing a wider 'own brand' range of family products will strengthen the Hasbro name outside its already wide range and promote Hasbro's child development credentials. This direction would promote learning through enjoyable challenges, where valuable life skills would be acquired. The product range includes fun toys that enhance individuality, whilst also allowing family members to get involved.

Cheeky Chimps
Future concept for Hasbro

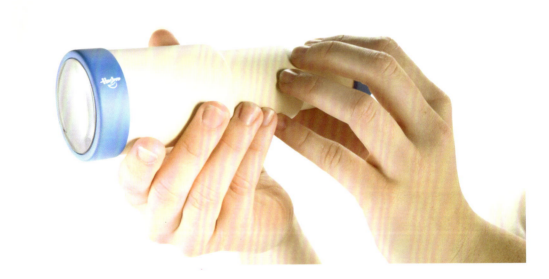

Cheeky Chimps transforms the traditional treasure hunt into an interactive game for both parents and children. These little monkeys can be hidden away in secret locations; only once they are found will they unlock the treasure that waits at the end of the search. Each explorer is armed with a telescope, looking through the lens informs the explorer when they are close to the chimps by the display changing colour.

Each telescope is connected to four chimps, with the aim of the game to catch your chimps first. Once all the chimps have been found, they collectively unlock the treasure. With new technology becoming more intrusive within children's lives, the nature of outdoor play can be lost. Cheeky Chimps encourages children to engage in outdoor exploration and increase their levels of physical activity.

Katie Henbest
Industrial Design and Technology

Laughter and fun are what personify Hasbro toys, making them one of the most successful toy manufacturers. Squizzy uses these principles and adapts them to tackle an ever-increasing issue. With children believing that they know what is best, establishing acceptable behavioural patterns can be a challenge for modern parents. Squizzy the squirrel aims to provide that helping hand to parents while giving children an interactive toy to care for. Squizzy is happiest when he has acorns in his log, and displays affection through fun swishes of his tail. This can only be achieved if the child collects enough acorns from their parents by achieving goals and set tasks. Squizzy helps parents to promote good moral values and behaviour in a fun and exciting way, which are increasingly important values as they grow older.

Olly Brown
Product Design

Peek-A-Penguin
Future concept for Hasbro

Hasbro encourages children to have fun through the use of new products. Peek-a-Penguin is an interactive Hide and Seek product that can be used in and around the home, encouraging inquisitive behaviour natural to all young children, and helps to develop a healthy imagination to prolong childhood and help to keep them younger for longer. Peek-a-Penguin combines the traditional games of 'Hide and Seek' and 'Hot and Cold'. Once switched on, the Penguins are hidden by a parent around the house in various places and the child uses the egg to locate each penguin. The egg glows hot (pink) if you are close or stays cold (icy blue) if you are not close to a penguin. The aim is to find all of the penguins so they can huddle around the egg and keep each other warm.

Charlotte Pharez
Product Design Engineering

Why can't homework be fun? Professor Feathers is a new product for Hasbro where children are able to learn whilst playing. Similar to a dance mat, the child has to jump on the correct answer using the different stepping blocks. The answers light up on the blocks and it is up to the child to find the right answer. This can also be played as a group, so learning can be a family activity. There are three different modes for spelling, counting, or learning shapes and forms. With this interactive game the child can get feedback with sounds and also on-screen. This toy is aimed at children between the ages of 5 and 10 years old with different levels of difficulty available.

Rosanna Wells
Product Design Engineering

Small Footprints
Future concept for Hasbro

Small Footprints provides a competitive platform for children to positively engage with more economical travel. The watch is worn by the child and monitors their travel throughout the day using GPS and heart rate signals. When detached and placed in the hand of the child's personalised character, the stands interact with one another and grow in height depending on the child's latest travel activity. Who can be most efficient? A fun arrangement of personal characters, Small Footprints encourages healthy competition between children whilst engaging them in one of the most critical environmental issues: travel and fuel use. Small Footprints aims to bring the whole family together and support children in the act of stewardship.

Tom Reader
Product Design

The theme inspiring this collection of concepts is the notion that Marshall customers are "Born with Attitude". The range is designed to be intergenerational, allowing young children to grow with the brand. Rock music has a distinctive and infectious sound that appeals to people of all ages, allowing parents to share their passion with their children, or younger children to collaborate with friends or older siblings. Products in the range are designed to appeal to different age groups; acting as stepping stones towards Marshall's world-renowned range of guitar amps, fueling their growing passion for Rock culture, keeping Marshall close to them throughout their lives.

My First Marshall

When a child sees music being played, they are intrigued. Music is stimulating and exciting, playful and descriptive. The act of playing an instrument, especially one that involves a lot of motion, like a guitar or the drums, is something that children want to emulate and participate in. Sharing these early experiences builds an incredible bond between parents and their children. Also, the adults will be enthusiastic; because they are doing something they love as well, sharing their passion with their children. This concept gives Key Stage 1 children aged 5-7 who lack the necessary skills and coordination to play the drums or guitar; the opportunity to express themselves creatively. It will be their first foray into "Marshall Law", laying the foundations for a relationship that could last a lifetime.

Dan Smith
Industrial Design and Technology

The Marshall DJ Companion bridges the gap between old school and technological DJs, an inter-generational development to the way that DJs create their music sets. The product aims to represent the traditional method of sorting through a record box to find the next song, reigniting the importance of the album artwork. The product is in the style of a flight case that splits into two angled halves, each mounted with touch screens, enabling the DJ to sort through song artwork and create an appropriate playlist to set the mood for any environment.

Lucy Kierans
Industrial Design and Technology

Marshall Micro Mixer

Future concept for Marshall Amplification

Reflecting Marshall's rock and roll roots, this inclusive handheld device will connect with old and young alike. To counteract a trend towards the passive appreciation of music, the Marshall Micro Mixer will appeal to an electronically orientated market by providing a product to hold and to play. The Marshall Micro Mixer utilises two complementary interfaces: the first is analogue reflecting the history of Marshall while the second projects a modern digital influence. The knobs and the touch pad encourage user engagement presenting tactile mediums that enable self expression. In this way, the Marshall Micro Mixer empowers all abilities to explore a musical world of their choice and of their design, to be a one man band or to interact with others to compose, record and to play...live and loud.

Max Latimer

Industrial Design and Technology

It is the simple things in life that are the most extraordinary and the tiniest things which help to inspire young ones, leading their imaginations. With today's ever increasing busy lifestyles it is important to keep track of your family's agenda. Reminisce with old photos, laugh at videos and listen to music from anywhere in your house. The JMR is a multifunctional device brought to you by Marshall, for the entire family to share and enjoy, young and old alike. With user definable settings, age is no obstacle. Easily control and play media on the remote or throughout the home wirelessly, with no fuss. It features a large HD touchscreen, whilst still boasting the renowned iconic tactility associated with the Marshall heritage. A revolutionary home media controller at your fingertips, ready to be turned up to 11.

Stuart Wickens
Industrial Design and Technology

Guitar Tutor

Future concept for Marshall Amplification

Many people are interested in learning to play a musical instrument, and are often keen at the start, but are easily distracted by early difficulties and due to this give up learning. The electric guitar is no exception to this, but this situation need be no longer, with the Guitar Tutor from Marshall. A known musical track is digitally decoded and transmitted to the electronic LED lit rail which is fitted to the top of the guitar's neck, the LEDs within this rail are positioned on top of the known positions on the guitars fret board. Once the music is received it is displayed on the lit rail in a corresponding colour which matches the string on the guitar allowing a note, chord, down pull or up pull to be played, all co-ordinated to suit the learner's ability.

Rob Wooldridge

Industrial Design and Technology

Marshall

BORN WITH ATTITUDE

Werther's Original is a brand built on traditional family values. As finding quality time for family becomes harder due to everyday obligations in the busy competitive future, the product concepts have been created to encourage family interaction and bridge the gap between generations. Not all memories last forever, but the moments these products make are precious to everyone involved. The products are timeless, both innovative and classic in execution and can be passed on to loved ones for many years to come. They offer a range of traditional activities that create moments through sharing, teaching, learning and interacting.

Werther's Shaver

Future concept for Werther's Original

The Werther's Shaver has been created to build memories from moments when a son watches his father shave. This product would be precious to the owner with the aim of it being passed down from generation to generation. The shaver and its box have been produced from the finest Teak wood, for its waterproof properties, and the box had been carefully crafted to resemble Werther's confectionary. The aim of the box is to intrigue the user to explore what is inside, and to differentiate it from other boxes so that the user feels special.

David Cole

Product Design

Werther's Recipes

Future concept for Werther's Original

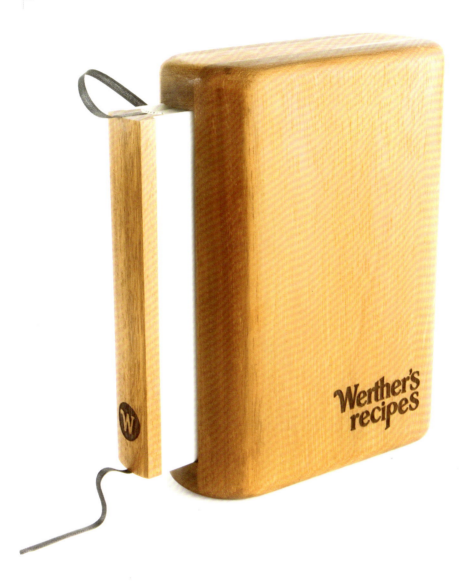

Werther's Recipes aims create memorable moments for all the family to share now and in the future. The wooden casing creates a home for all the treasured recipes passed down through generations of the family. Combining old and new, past and present memories can live on inside the Werther's recipe book. The family can share experiences by spending quality time in the kitchen, recreating old and new recipes, and experiencing the tastes and treasures.

Amelia di Palma

Industrial Design and Technology

Werther's Chess Board

Future concept for Werther's Original

The basis for the Werther's chess board stemmed from the idea of increasing family interactions and bridging the gap between generations through different activities. Chess is a hugely popular game worldwide, culturally long established, appealing to everyone. It is a game which is taught and passed on through generations, fitting in perfectly with the Werther's brand's core values. The chess board itself was also designed to be a display piece aimed to attract and interest people to want to explore and play with it. Its smooth, soft and natural forms highlight the relationship between the game and the brand.

Mohamed Abdel-Gadir
Product Design Engineering

A product that not only brings all generations of family together, but also keeps the memory of that family remembered for the future generations to enjoy. The Werther's Capsule allows all family members to work together, gathering treasured memories and valuables, reminiscing as they choose items for future generations will appreciate and enjoy. A day out can be planned for all the family to take pleasure in, and a place of sentimental value is chosen for the burial of their capsule(s). Each family member who takes ownership of a Werther's Capsule will have the chance of selecting a special piece of Werther's Jewellery. In addition, the jewellery uses Bluetooth technology to pair with the capsule when near, vibrating more frequently until directly at the location of the treasured memories.

Jayson Tulloch
Product Design

Stabiliser Wheel
Future concept for Werther's Original

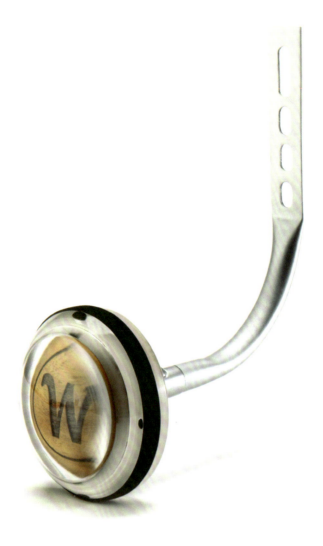

The Werther's stabiliser wheel provides support when learning how to ride a bicycle. We all remember the first time we learnt how to ride a bike and who taught us. This product represents the gathering of loved ones experiencing those moments that we will always remember, those Werther's moments.

Yinko Branco - Rhodes
Industrial Design and Technology

Werther's Original

your moments, your memories...

Christmas Present

A Piece of Creative Writing

Well, it was all my idea. Christmas always makes me a bit soppy. Nostalgia overload I suppose.

So, this particular year we were sitting round the tree with a glass of mulled wine, that wonderful warm spicy drink, as good as a lover and nothing to stop you having more when you want it. The grand-kids were playing with their new toys and I was looking at Moira our daughter and Pete her patient husband who puts up with us all. Mark, always so mellow, was putting on the carols CD.

Then there was my dad Stan, grandad to everyone, great grandad to Ethan and Sallie, who's been at all my Christmases since I was born. Still missing mum, still smoking his pipe and drinking our health in his favourite sherry. Still eating his favourite butter candies even though they must have demolished a few molars in his time.

I know what you're thinking – pass the sick bag – but I thought how nice it would be to capture the moment. Not in a photo, but something tangible that would bring back the Christmas of 2005.

Reckon we're all too good to be true? We've had our difficult times, such as when Mark lost his job (there were few presents that year and turkey burgers for dinner) and of course when I lost my mum that December. That Christmas was a blur.

Now everyone was settled and I thought: if only you could make time stand still. So that's when I suggested burying a festive time capsule, to be opened in five years time. I ignored the grumbling chorus of "Oh mum...". At least the grand-kids at four and eight were keen.

Now I'm not so sure it was a good idea. It's time to look again at our capsule. Dad's no longer with us – a heart attack last year just before his 94th birthday - and I can't imagine Christmas without him. Ethan's nine and likes to bury his head in computer games. Sallie's now a teenager and doesn't want to be with us at all. Mark's worried about the lack of interest in our bank account and is not so mellow. I'm just plain tired all the time.

Nevertheless we all gather to empty the 2005 capsule which has been 'buried' in the loft for half a decade. First up is Ethan, who is aghast at the evidence of his four-year-old taste. There is a picture of his Robosapien (obviously no one was expected to put an actual present in the capsule), which was then the latest high-tech robot, a book cover of The Very Hungry Caterpillar and a packet of Percy Pig sweets.

Sallie stops laughing when she sees the wrapping from her Bratz funky makeover styling head and we all wail when we see her pick 'n' mix from Woolies. It's now defunct, and we miss it.

The adults' choices are predictable: from a candle still giving off a spicy smell, to that year's Christmas edition of the Radio Times. The news pages from five years ago are filled with wars and flu epidemics. No change there.

But it is grandad's choices that really get to us. We all stare at his spare pipe, a miniature bottle of Bristol Cream, and a packet of his boiled sweets. These items, so simple and ordinary, had symbolised for him so many happy family Christmases.

Even Sallie melts. "Let's make a toast to Great Grandad," she says, reaching for her can of Coke as Mark goes off to open a bottle of red.

"No, I've got a better idea," says Ethan, "let's toast him with his favourite things in all the world." He takes a Werther's Original from the box and hands them round. We all solemnly unwrap a sweet and toast my dad as we pop them in our mouths. Suddenly we feel like a family again.

"This is what comfort tastes like," says Ethan, remembering the slogan for the advert for Grandad's sweets, and we all laugh.

By Barbara Fisher © 2011
Creative and Professional Writing MA
barbarajoycefisher@hotmail.com

Airfix is a very specialised brand with a loyal but relatively small consumer base. Due to its strong brand values and continued focus on its customers, it has survived in today's ever growing technology-driven world. Attention to detail and the capturing of memories are key to all Airfix products, with existing models carefully replicating old war vehicles and figurines. With such a niche product range, the future for Airfix is uncertain. These future concepts harness the idea of the user 'building memories' to share with future generations. They include both proactively creating new memories and also providing products with which the user can store and share already existing memories.

Airfix Sculptors
Future concept for Airfix

A game aimed at adults which will add an extra dimension to an evening in. The kit is based around modelling clay which cures in the air in 24 hours. This will allow the users to produce lasting models. The players will produce one model each over the course of the night based around a subject selected at the beginning of the night. An example of a subject would be to make a model of an object representing the first time you met the person on your left. The game will come with creation tools including an extruder, rolling pin, moulding boards and an interchangeable sculpting handle. The model can be given to the person who it is about, they will then have something which is very personal to them and will be a modelled memory.

Tom Collett
Industrial Design and Technology

Everybody, whoever they are, has a sentimental memory with a best friend, whether it is a ticket from a movie, or a rock from the seaside. These memories are precious and should be admired. Airfix has strong brand values relating to personal memories and the sense of building and construction. This future concept for Airfix allows you to showcase your fondest and most precious memories in your own personalised display box. Self-constructed magnetic windows can be assembled to create any shaped box with no limit to the user's imagination. Conservative users may follow the construction examples provided but, for more creative individuals, the assembly possibilities are endless. The box can be reconstructed as many times as you wish, so the box can adapt to ever changing memories.

Alex Williamson
Product Design Engineering

Airfix Run

Future concept for Airfix

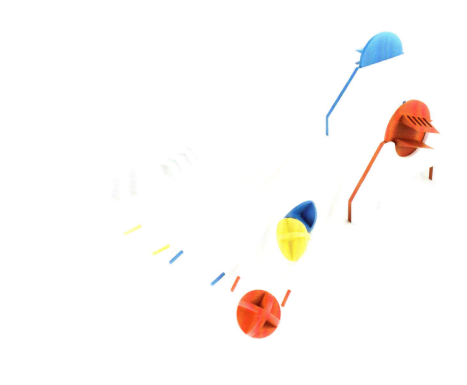

The Airfix Run is a cool, creative construction set. Targeted at young teenagers the product provides the basis for an elaborate domino run. The pivots allow the user to create their own catapult, combined with interconnecting slides, dominos and balls. The Airfix Run enables them to do something creative and rewarding. The result is as resourceful and imaginative as you like and is a fantastic way to spend a Saturday afternoon; why not video the result and show your friends?

Sam Verma

Product Design

Today, a family can have many connections such as step families and adopted siblings. With this being a common case, a family relies on building bonds, to create lasting memories that bring them closer together. Step and adopted siblings become part of a family at different stages of a family's life. There needs to be a way to secure these memories and make every member feel equal and involved. The Airfix Wee Mee is a personalised figure of a family member that is then placed on a central "Home" to acknowledge when a family member is present in the home. Every individual in the family personalises their own Wee Mee, and carries it with them when they leave. When they return, it is then placed back on the central "Home". This allows for the inclusion of all the family members and helps to strengthen the family bond.

Jamie Trigg
Industrial Design and Technology

Airfix Build-Your-Weld

Future concept for Airfix

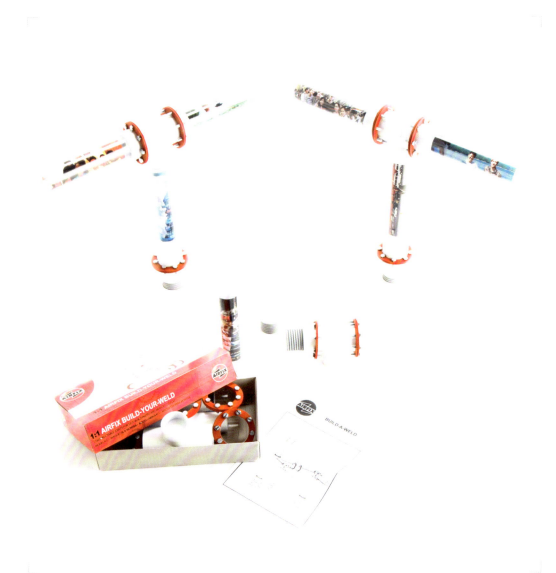

Airfix is one of the UK's most famous traditional companies. Most people know Airfix whether or not they have bought into the brand and there will be a great number of older people who can remember many good times building and constructing the toys from their assembly kits. It is this sentimental value that was central to the design development of this project. The idea is to allow the user to display their photos and memories in their own unique way through the provision of constructible parts leading to a different set-up each time. The focus is on building memories and in allowing the user to build their own format for displaying their memories, this design really integrates the key core values of Airfix.

Michael Day

Product Design Engineering

A new way of making your mark, on every object surrounding you. The concept was developed as part of the new direction for Airfix; 'Building Memories'. You can engrave your initials and create your memories on a piece of rock, a plate, a mug or even a tree, and your personal mark will last forever. The Power Brush concept contains a mini air compressor motor, spinning at 800 rpm, giving you the most powerful engraver driven by a single battery. The form is ergonomically designed to give you that added precision and control. A single touch operating button activates the angled head. This unique engraver is not a workshop tool, but a handy product to embellish everything around you.

Mohammed Elsouri
Product Design

ProDriver
Social responsibility in driving

MultiCycle
A safe multifunctional cycle for mothers and children

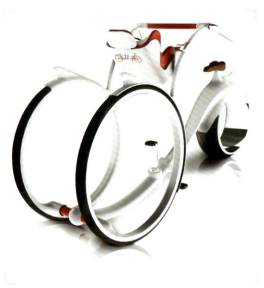

With increasing fuel prices and ever more stringent laws for driver and passenger safety, it is a good time to use design intervention to change driving behavior. ProDriver aims to reward drivers for good driving behaviour.

The service-based product is essentially a smart phone application. Using the inbuilt camera, GPS and accelerometer, it is possible to detect how safe and economical a driver is being. This information is displayed in a clean ecological interface, which improves a driver's situational awareness of the road.

When driving, the user can access the system to detect which zone they are traveling in. The green zone means that they are driving in a safe and economical style, and in return for every mile driven in the green zone, the user receives one ProMile. When the driver is in the red zone, they are driving in either an unsafe or uneconomical style, and in return ProMiles are deducted from whatever they have accumulated. There is also an amber zone, which is in between the green and red zones. When in the amber zone, no ProMiles are either gained or deducted. ProDriver aims to give drivers a nudge to behave in the correct style.

The ProMiles are then banked to a user's individual profile in the ProMile cloud. These are then used to buy discounted fuel and insurance and to compete with their friends and colleagues.

Multi-Cycle is an eco-friendly adaptable cycle for the busy city. Research from TFL has shown that there is a rapidly growing trend for cyclists in the city. In addition, it has been found that mothers are unlikely to cycle with their infants due to safety implications. Multi-Cycle is aimed at encouraging cycling in congested cities and to change cycling behaviour towards mothers.

It is adaptable for different commuting needs; from a three-wheeled bike with a child seat to a pram, or even, a simple narrow bike. Safety has been a key element throughout the design, and has been applied in order to boost the confidence, and safety perception for the user. Moreover, this will be the first bike with active airbag systems that will protect the child even in a dangerous collision. Extra attachments also allow the bike to be used by parents and children of different shapes and sizes. Another positive feature of the bike is, it can be folded together hence less storage space is needed, making it perfect for compact cities. MultiCycle – the safest eco-friendly child transporter.

Premal Mistry

Arber Shabani
Integrated Product Design

The final design, Curve, is a simple and elegant solution to the pain and discomfort most female cyclists currently experience with standard saddle designs. Curve's unique shape compliments the shape of the female pelvis and support her weight evenly. The torsion springs allow the saddle to tilt forward when there is an increase of pressure at the front of the saddle. This tilting feature reduces the amount of pressure that the cyclist experiences on the pubic rami at the front of the pelvis. Three different spring stiffnesses of spring and sizes of saddle mean that a variety of female body shapes can use curve effectively and experience its benefits.

Katy Koren
Product Design

Madeleine Carver

This is a Man's World

The world of women and work can be a sensitive subject, normally tackled by the militant feminists or chauvinistic males, but it is still one that deserves attention. The design world is heavily saturated with testosterone and when asked to name significant female product designers, most people will just give you a glassy-eyed stare. My question is: why in this profession is there such a persistent lack of female presence and what are the reasons behind it?

In 1985 Margaret Bruce said that things were never going to change, as "this is a man's world and always will be", yet this is not the case. At the time, women made up less than 1% of industrial designers. Many problems faced them in the design world; it was often thought that women did not have the skills to become designers. The perception was that women were better suited to 'styling', 'colour' and 'co-ordination'. Things have changed greatly, as boundaries are blurred between the traditional roles of men and women.

Men often still earn more than their female counterparts; but progress is being made as the gap between wages is narrowing. Things have come a long way from the Equal Pay Act of 1970. Jane Atkinson (of Project Associates UK Ltd) believes that the genders already have similar pay and status and it is only at the top that the pay between men and women changes. She puts this disparity down to a woman's choice between a career or a family. For most women, by the time they are starting to build a career it is time to have children. When this happens, women are faced with the tough decision of whether to step back from their careers or to try balance the two and risk missing important milestones, aware that their children's early lives can never be repeated. The shift in roles is not just about women making their way into the workplace, but men finding their way into domestic roles. Future legislative changes regarding paternity leave may not only enable men to have a greater role in the first year of their child's life, but also allow women to choose to return to work sooner, as the duty of care becomes more balanced.

When hiring, women in their mid-twenties to early thirties have the potential to be by-passed as they are of child-bearing age and employers may see it as a risk. Realistically, it is difficult to keep a job open for an employee who will be absent for at least 39 weeks. Having staff members with the commitment of children is a lot less appealing than those who can be more flexible in their hours.

However, gender bias can work both ways. Positive female discrimination also exists, for example through the use of gender quotas. These are used to artificially create a balance in a naturally gender biased situation. A positive aspect of gender quotas may be as guidance rather than legislation, as they can ensure a healthy balance to encourage both variety and diversity. If it were to become part of legislation in the UK, then it is doubtful it would increase numbers of women in design, as more needs to be done to add to the appeal industrial of design. The application of gender quotas in the workplace must be discouraged, as they can be demeaning to both genders. If rigorously upheld, quotas are stifling, and risk someone losing out on a job to another less skilled or talented.

On 'Speak Up', a graphic design discussion blog, it was noted that female designers are less vocal with their opinions than their counterparts and do not garner their potential recognition. We have to make ourselves known by putting ourselves out there. As women, we cannot demand that diversity and equality just be given, we must create it for ourselves.

Looking at the roles in industry, women typically gravitate towards more humanistic design roles such as research and dominate other design professions such as the design craft world. Stereotypically, women prefer the human-centred side of design. Women often seem to enjoy design that is more hands-on creatively, which could be put down to the belief that women use design as an opportunity to be creative rather than face the fast-paced, profit-orientated industrial design world.

A lot of our beliefs about both genders are based upon patterns and trends, and while not always fair they do usually reflect the majority. Research shows that if the same group complete two questionnaires, the results of one that asks participants to reveal their gender will produce results that conform more strongly to gender stereotypes, which may be driven by our perception of others' expectations. This harks back to the nature vs. nurture debate: is it inherent in our genetics for men and women to be different, or is it the way we are brought up? Physically we are undeniably different, in terms of our mental capacity is there really a difference?

Those that advocate nurture as the main influence believe that the "neurotrash industry" tries to make the old-fashioned stereotyping plausible, but neurologically, there is little difference. While there are structural differences between the sizes of male and female brains, there is no proven link between this and reduced functionality. From birth we are designated colours according to our gender and with constant media bombardment it would be impossible to raise a child in a gender-neutral environment. For supporters of the nature argument, the belief is inherent differences between men and women evolved from our hunter-gatherer days where different hormonal balances and brain structures were needed. Traits such as lack of spatial awareness, and a good ability to empathise and register emotions are part of what differentiate women and men. It seems a combination of both nature and nurture create the gender differences, but it would be impossible to prove this.

There is currently a growing trend for stereotypically 'female' characteristics being more valued in the workplace. This can be linked to a change occurring in the design world, as industrial design becomes less technical and more about the user experience and intuitive responses. This is something that appeals much more to women both as designers and consumers. As consumers, women are found to be up to 80% more likely to make or influence a household product purchase decision, which is starting to change product design. No longer happy to accept pink and floral 'gender slapping', women now demand more sophisticated products that not only appeal visually but are more intuitive in their use. 'What Women Want' is becoming a lot more important as a result and this is causing changes in the way we design products. This change could lead to an influx of more female designers as they know the answer to the question.

The world of design is incredibly competitive, and it is still very much a man's game. The only way for women to really become more visibly successful in the design world is for a change in the design process to occur and for them to become more vocal about their achievements. Any existing problems faced are ones we make ourselves, and when women create up-cries about the injustices that befoul us, they conveniently seem to forget that gender bias works both ways and we are also often positively discriminated towards. Society may have had a structure to it in the past, but things are changing every day, and it is up to women to help enable these changes and make an impact on the design world.

Dettol Sense
Hand sanitisation on the London Underground

With the practice of good hand hygiene on the London Underground having a vast effect on the absenteeism of London's commuting public, Dettol Sense presents a solution to improve the health of the passengers on the tube. The 6.5ft sanitation station, offers up to four commuters at a time the opportunity to sanitise their hands, eradicating bacteria transferred onto the skin during their journey. A hands-free solution has been developed to incorporate the trustworthy brand recognition of Dettol into the fast-paced environment of a London tube station.

Lucy Kierans
Industrial Design and Technology

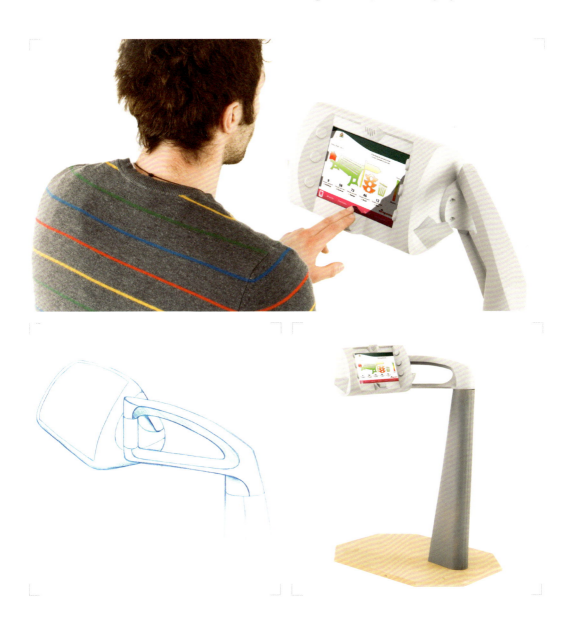

Voice Your View is a system harnessing people's feedback for the ongoing development of safer and more fulfilling public spaces, in collaboration with DesignPlus. The device is used in public settings providing more convenient and engaging access to existing feedback channels. Local authorities and planners gain increased public involvement in key decisions. This product capitalises on existing behaviour patterns, allowing users to browse other comments and local information in a relaxed environment. The device can relay commercial messages on one display whilst engaging the user with a more personal interactive experience on the other. Designed for accessibility and simplicity of use it captures audio feedback and deploys gestural and tactile input to appeal to a wide audience.

Jermaine Legg
Industrial Design and Technology

Zonk
Portable baby calmer

Zonk is a device for use by parents and their young children (0-1 years) in public scenarios, that allows them to calm a baby that has just started crying or who is about to cry. It works using the mother's proximity and vibration to calm the infant. Zonk uses two harnesses: one for the parent and one for the child. When the child is about to cry, the parent picks them up, slots them onto their own harness so the baby feels near to them and can be comforted. For further comfort they can activate the switch located on the top of the waist strap in the baby's harness to turn on the vibration, which is localized to the baby's waist strap. Both harnesses are designed for optimum ergonomic positioning and comfort.

Timothy Dunkley
Product Design

This public terminal will allow people to communicate easily about the public spaces around where they live. This will help improve the local area and quality of life because local authorities will be able to act on the suggestions made to improve the environment. This terminal will be situated in a supermarket, offering safety for the product and a more comfortable environment for the user as well as offering a friendlier face to the supermarket. This could be used to help supermarkets show that they are working and engaging with the local communities they are located in, creating a positive brand image. The terminal will be situated in both in and out of town supermarkets so that the terminal is in reach of as many local people as possible creating a community voice.

Fred Swallow
Industrial Design and Technology

Man's Best Friend
Reducing dog fouling in public areas

Man's Best Friend is a product designed to improve the issue of dog fouling in society today. The product focuses on the picking-up process that is usually carried out by the dog owner. The product accompanies a campaign to increase awareness and highlight the serious dangers and consequences that come with dog fouling in public areas. Man's Best Friend is a two part product comprising of a pick-up glove and specially designed pick-up dog poo bags.

Hannah Haile
Industrial Design and Technology

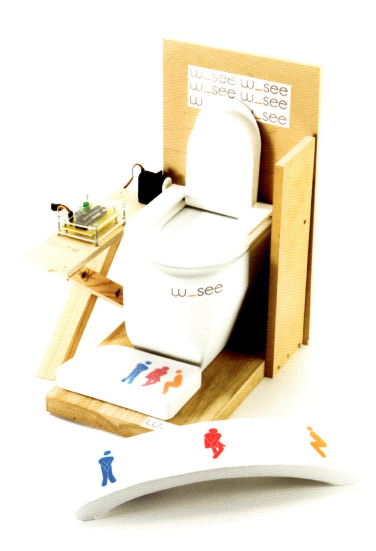

W_See is a new toilet system aimed at reducing disease transmission within the restroom environment. With 40% of public toilets in the United Kingdom being closed in the last decade due to lack of cleaning and sanitation, we run the risk of losing further facilities if this problem is not addressed. The toilet system is divided into three parts: the seat, the electronic circuitry and a touchscreen LCD interface. These aspects allow the user to enter the toilet cubicle, select their requirements, use the commode and the technology will do the rest to leave it hygienic and ready for the next user. The product uses motors to lift the toilet seat and lid up and down to suit the different needs of the toilet user.

Nicholas Edgar

Industrial Design and Technology

Beat the Burn

Sun protection device and campaign aimed at parents in the UK

Around 11,000 cases of skin cancer are diagnosed in the UK every year, and this is estimated to rise in the next 15 years. Children are some of the most prone to sun damage, and a clear correlation has been noticed between over exposure at a young age and skin cancer in a later life. This campaign is targeted at young children and their parents, encouraging them to behave more responsibly in the sun. This concept involves actively reminding the consumer that there is a constant danger, using more fun and exciting methods. The colour changing sunglasses are the main aspect of the campaign, allowing the user to take a reminder with them that will not be intrusive into their day-to-day lives.

Bradley Wherry

Industrial Design and Technology

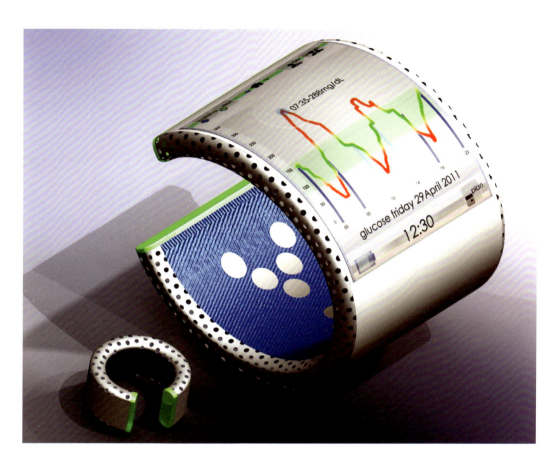

With an ageing population the number of patients with long term diseases and resulting health issues is steadily rising, increasing the strain on the NHS. Conditions such as diabetes, high blood pressure, high cholesterol and obesity are closely linked with cardiovascular diseases, the leading cause of death in the UK.

This innovative device proactively supports health self-management and improves the user's general health through the provision of tailored information, peer and professional support and lifestyle incentives.

The bracelet-ring checks and monitors health data such as glucose levels and blood pressure, and keeps track of food consumption and exercise progress. Other monitors such as weight scales can be directly connected to the bracelet via radio signal. The measures and trends can be seen at any time by the user and health professionals.

This ergonomic, aesthetic and non intrusive solution offers the user an intuitive way to keep informed about their health status as well as easily receive the needed information and support with one touch. The device supports savings for the NHS with a positive answer for patients as it helps shift understanding from clinicians to patients.

Social and physical inclusion were taken into account throughout the project. Activities were carried out with the REACTION (European level research programme) programme to ensure the project fits in with future plans for the health care system and patients and clinicians were actively involved in the design process.

Tilia Van Olmen
Integrated Product Design

Chris Tarling

Our shared journey so far...

From Shoreditch campus to the busy streets of New York City, my Brunel course gave me the best grounding for an exciting career in design.

I have now lived and worked in Manhatten for nearing two years. My employer, HCT Packaging Inc, gave me an opportunity of a lifetime: I had to start an ever-expanding design force on the east coast of America and be based out of the showroom in Rockefeller Center.

I now design primary packaging and beauty tools for the world's leading cosmetics companies. Some of the clients I have worked for recently are Bobbi Brown, MAC, Laura Mercier, L'Oreal and Victoria's Secret. This usually consists of a customer briefing us on a new programs two years ahead of its predicted launch and creating visuals or models for presentations.

Innovation in my industry comes in many different forms. It can be an amazing Italian formula, or futuristic materials, a distinct dispensing device, or new electronics . I am sure many design students hear the term "K.I.S.S." along their journey at Brunel and innovation can also be found in the simplest things.

My design journey began with model making for OTO3D in Surrey where I actually used to create items for two of my future employers. Model making skills gave me a good eye for detail, which benefited me later in NY when creating new conceptual makeup ranges for clients such as Calvin Klein, Donna Karen, Côty and Estee Lauder.

I then progressed onto running 100 tonne injection moulding, blow moulding and laser sintering machines at Procter and Gamble. I would use CNC skills to create tools, optimise machines for different plastics and then run single injection or over moulding for pre-production samples. Working for such a large corporation enables you to learn a lot from a large skill base of talented people. I definitely owe the "core" creative team huge thanks for helping me learn new materials, machining techniques and strong corporate skills. During my time at P&G I also trained new Brunel work experience students, which was a great pleasure and good to see them coming through just as strong.

HCT captured my attention with its industry leading innovations and customer base, which was second to none in my opinion. Having the skill set from Brunel and the experience I had acquired since leaving university meant I fitted in perfectly with the design / engineering teams straight away. My global design director, Rebecca Goswell, was, and still is, a true inspiration for trends and design thinking. I had worked in the UK office for just over a year when I was approached to take the NY position. If anyone ever tries to get a US VISA, especially a good one, then be prepared for a long process, but the interviews are fine and once I was here I hit the ground running.

Since I landed in NY and took over the company's east coast showroom I honestly feel like I have not stopped. Riding the subway, walking from east village where I live up to work, or generally getting lost in this amazing city are just a few things that inspire new concepts for me. My usual week consists of presenting innovations to a range of customers either specific to their chosen areas or sometimes to showcase "our latest and greatest". When the briefs come in from customers I timetable design time for NY and communicate with my colleagues in our two other design offices in the UK and California. This is so that high profile work can be shared and given the global dedication that it warrants. Sometimes smaller localised briefs come in and my coworker Ashleigh Earl and myself will deal with them just on the east coast. This enables faster turnaround time and more face-to-face interaction with the customer to meet specific needs.

We recently worked on high a profile global design project for a cosmetics show in Italy called "Cosmoprof Bologna 2011". Our two senior designers from the UK are Tom Bentley (a fellow Brunel graduate) and Federica Medina. They used their incredible expertise to create some never before seen innovations in nail, face and eye categories. Altogether we created a very impressive range of concepts from London, NY and LA, which I am pleased to say from first hand experience left our customers suitably impressed.

Designing for east coast lead brands comes with amazing opportunities to take a design and launch a year later for over 1 million pieces in the US alone. My first ever fully designed range is for Bobbi Brown and will be used for her holiday ranges over the next two years. I have been very lucky to have been given the chance to meet directly with many of the leading cosmetics companies and take design direction straight to our teams for better design co-ordination.

I am not too sure how many people can say this, but I love my job with a passion and a drive. This is due to the people I work with, 100%. Without designers helping globally, the engineers providing incredible tools and also project managers keeping us on time we would never have become such an amazing company in my opinion. The only piece of advice I would give to anyone is that anything is possible: just take a chance and go for it. Your friends and family both at Brunel and at home will always be there for you...so what are you waiting for?

"It's only make up"

Francois Nars

Chris Tarling, Area Manager, HCT Packaging Inc.
Graduated in 2005

Designing Spaces for Designers

Spatial design strategy of physical work environment for creatives

Work environments have always been a critical issue in terms of performance and profit of businesses. Many researchers have revealed that physical environments improve employees' work performance. Recent research also shows that creativity plays an important role between physical environment and work performance. Physical work environments enhance employees' creative thinking and creative thinking improves work performance. However, every organisation cannot use the same space scheme to enhance employees' creative thinking. According to Meel, Martens, and Ree, different organisations need different work environments based on their own work processes. Existing models address general offices regarding creativity, but not for creative industries or designers that have their own unique work processes.

This research explores how physical work environments can improve the creativity of designers within creative industries. For the purpose of this research, creative industries in this study are restricted to client-based businesses that need to think in different ways to develop innovative solutions including design, branding, and advertising. A spatial work environment design strategy will be created for designers as a result of this research through examining designers' work processes. Several research methods will be used to create the strategy such as case studies, interviews, observation, and detailed 'office' surveys. As a key design research tool, visual observation and audit will be critical for this research. Unstructured workplace interviews will also help to give colour and insight on current problems of the contemporary work process or 'journey'.

So far this research has found that organisations need to invest in physical work environments and there is a clear need to develop a spatial strategy in order to enhance designers' creative thinking. Latest research focuses on relationships between physical environments and social interactions while former research has been restricted to physical interactions regarding work environments. Also, corporate culture has risen as a key factor that interacts with creative thinking within businesses.

These research trends mean that there is now a need to consider interrelationships among physical environments, individuals, and social communications beyond simple physical interactions. Consequently, this research will consider how physical environments can make a creative corporate culture that encourages designers' creative thinking. It is anticipated that the final result will address both physical and social human interactions in a comprehensive perspective.

Hyein Lee
Design Strategy and Innovation

The design process.

A compelling argument already exists for why effective use of design can bring enormous value to organisations. Research from UK's Design council confirms that companies who are 'design led' are out-performing those which do not use design effectively. However, there are profound differences between the working methods and processes of designers and conventional business managers. For example, as Liedtka says, "Business demands prove we've arrived at the correct answer" while in contrast, design has "a penchant for doing rather than carrying out extensive prior planning" and has a very experimental approach. Furthermore, design is fundamentally very difficult to quantify. This goes against the control and measurability valued by most business executives. There is also evidence that design is largely misunderstood by non-designers to mean styling, rather than 'strategic problem solving' as the design industry would prefer it to be known. The consequences are poor multidisciplinary communication, which is a barrier to creativity and innovation.While the design profession is making in-roads to solving these problems by improving the business skills of designers, this is not enough to bridge the gap between designers and non-designers. The initial findings of the research reveal that design is almost completely absent from business education, and features as a small and often optional part of MBAs. Moreover,

according to Hollins, "when it is offered as an optional module, the take-up tends to be low... students are not convinced of the usefulness of the topic". This leads to the key question of the research project: "How can non-designers in organisations be given a good understanding of design?" The outcome of this research will be to create guidelines for effectively educating non-designers the methods and benefits of design. This research will first define what it means to have a 'good understanding' of design. This will be achieved through a series of expert interviews with design and business managers. The Design Management Institute's 'Value of Design' EU conference will provide valuable insights from both inside and outside the design industry, including discussions on 'Business Manager's Perceptions of Design Value'. This will help give a clear understanding of how design is currently perceived outside of the design profession, and identify benefits that could be gained by improving understanding. Some companies have schemes in place to educate their non-designers about design. It's expected that the final guidelines resulting from this research will recommend sustained training for non-designers in organisations, rather than a 'crash course' approach. It may also require a hands-on approach, where employees learn by 'becoming designers.'

Benjamin Kirk
Design Strategy and Innovation

Brand Stories, Communication and the Role of Design Thinking

What are brands and why do brands remain so popular? Wally Olins at a lecture at Brunel University gave a simple answer to that: "brands are the cultures people want to belong to". Moreover, Michael Margolis the President of Get Storied.com and founder of THIRSTY-FISH, one of the world's first storytelling consultancies, explains in his book, 'Believe Me', that people do not buy the products or ideas of brands but the stories attached to those brands. Laurence Vincent said in his book, 'Legendary brands', that a brand becomes legendary only if it can communicate its stories successfully to its consumers. So far it's clear that every brand needs a story and it has to be told well to consumers. But does it happen with every brand all the time? Matt Haig's book, 'Brand Royalty' shares many successful stories of great brands. There are innumerable popular brand stories in the market, but the salient fact prevailed on surface through the review of recent brand stories on BBC.com and in another book by Matt Haig, 'Brand Failures', that there are as many negative brand stories as positive. For example KFC have been facing PETA chicken cruelty allegations for over a decade now (kentuckyfriedcruelty.com) and is still losing valuable consumers worldwide because of that. If we follow Plato's great belief and Laurence Vincent's experience, then a way to overcome a negative story is to tell consumers the better, believable and favourable story. How can brands create favourable stories and how can these stories be effectively communicated to the consumers? 'Communication Models', by Denis McQuail and Sven Windahl, describes many models through which a message could be delivered successfully. However, none of them is appropriate for brand's stories. Even though brands communicate through mass media, models designed for mass communication could not be applied to brand stories because brands never convey generalised messages rather they always communicate with the consumer on an inter-personal level. There is no suitable communication for brands to create fruitful favourable stories. The second stage of research will use tools like comparative case studies, story boarding, expert opinions, consumer surveys and focus group to amalgamate stages of communication and design processes to develop new communication model based on design thinking. This new communication model will be suitable for brands to create better and profitable stories.

> "Those who tell the stories rule society." Plato

Ashima Amar

Design and Branding Strategy

One of the most significant trends that has been identified through research into branding today is the increasing influence of the consumer on brand behaviour. In particular, with the rise in popularity and use of social media websites such as Twitter and Facebook, consumers are more aware than ever before of what brands are doing. They are not afraid to make a judgement to the world if they feel a brand is misbehaving in some way. In a recent lecture Wally Olins, Chairman of Saffron Brand Consultants, further confirmed this thinking: he stated that "the customer is in charge" when referring to the consumer brand relationship. One consumer's opinion can be read and discussed amongst hundreds or thousands of individuals in a short space of time. This can be detrimental to a brand's reputation and difficult to repair. The consumer here refers to anyone who judges a company on some level based on reputation and could equally refer to employees or other stakeholders. The 'prevention is better than cure' approach is the way to deal with this trend. The need for increased transparency between consumers and brands is becoming apparent. Brands need to adapt to this trend. They need to more accurately gauge the effectiveness of their communications with their audience. In order to do so, they need to have the most appropriate and responsive tools at their disposal. The aim of the project is to create a design led brand audit tool that will facilitate this significant trend. Tim Brown of IDEO states: "Design thinking can be described as a discipline that uses the designer's sensibility and methods to match people's needs." The research so far indicates that many current brand audit tools are heavily marketing or business focused and although these factors are important, their effect on consumer perception should be deeply considered. Brands that achieve effective and consistent transparency in their communication are far more likely to succeed in today's market place. In the book "Zag: The Number One Strategy of High-performance Brands", Marty Neumeier states that "Customers today don't like to be sold, they like to buy, and they tend to buy in tribes" These tribes refer to peoples' need to buy into something which they believe in and are comfortable with. This sense of 'belonging' can be achieved by brands that focus more time and resources on appropriate communication with their consumer.

> "In the future the process is likely to be far more two-way, rather than brands imposing their view of the world."
> Simon Tempest, ACCO Brands

Jennifer McCormack
Design and Branding Strategy

Branding in the Public Sector

Developing a brand identity for the Bucharest public transport service

It is now well known what branding and design can do for a private sector organization. It is all about creating trust, about good communication of value, about long term relationships that can benefit both the producer and the consumer, about coherence, guidance, efficiency and evolution. But all these are also relevant when talking about the public sector. And while everything has been done or tried in the private sector, the public sector is still a blank canvas that remains to be explored by branding and design specialists. In support of the idea that public sector is the next big thing in branding and design, Wally Olins explained in the presentation he gave at Brunel University in March 2011: "It is hard, but it definitely has to be tried. You'll not get 100% success, but 60% is better than no success." The problem identified in the preliminary stage of this research project is that the Bucharest public transport suffers from a great lack of communication with the citizens. There is no trust, no efficiency, a bad management and therefore a bad image. Three

main opportunities arise from this issue as a possible outcome: to grow the awareness of public transport and thus grow the usage of public transportation and decrease city's traffic congestion (functional); to make the public transport a tool for city branding (emotional); and to make the people be part of the brand thus creating a proud and loyal community around the transport service brand (social). Therefore, the question that needs to be answered is: how can design and brand strategy create a better relationship between citizens and the public services? The research methodology focuses on qualitative methods: ethnography, focus group, customer journey, personas and scenarios. These are all meant to help understand the perceptions of probably the most passionate customers: the public services 'customers'. Main findings show that the lack of efficiency and communication in the management of the Bucharest public transport can be solved by design and branding. There is a clear opportunity to transform the public transport service into a brand and service that users positively respond to. The two issues that this research sets on to identify and solve are: how to evaluate the ROI of branding the public transport service and how to gain citizens' trust in order to create something equally profitable for the city and the people.

> "It is hard, but it definitely has to be tried. You'll not get 100% success, but 60% is better than no success."
> Wally Olins, CEE at Saffron Brand Consultants

Andra Mihaela Oprisan
Design and Branding Strategy

Every place, even if it is not branded, creates a perception in people's mind through associations and current opinions about the place. Romania's Capital, Bucharest, is the country's biggest city, industrial and commercial centre and Europe's 7th city with a population of 1,950,000 inhabitants. Bucharest is currently facing common problems that all the former Communist capitals are tackling relating to a state of social disorder known as 'transition' and a lack of economic sustainability. The reason for choosing this topic is to shift away from the negative perceptions associated with Bucharest, which is suffering from a poor image among internal and external constituents and unfavorable perceptions associated with it, often meaning poor or inferior. Beneficiaries of the project include the community, the external visitors and the stakeholders involved in Branding Bucharest. Furthermore, the key texts that have guided the train of thought in the research represent the caterpillars of the city

branding field; the core theories that embody the basis of this field's development, such as Simon's Anholt 'Brand New', 'City Branding: Theory and Cases' by Keith Dinnie and Michalis Kavaratzis' scientific research papers. One of the most inspiring and suitable quotes that could apply to a city like Bucharest was given by Keith Dinnie: "Don't be afraid to borrow from the best, while trying it your own way." The most important findings to date convey the fact that cities are neither products nor corporations and, therefore, a distinct form of branding is needed. Therefore, 'City branding is suggested as the appropriate way to describe and implement city marketing', (Kavaratzis), and 'the object of city marketing is the city's image, which, in turn, is the starting point for developing the city's brand' (Dinnie). The city brand audit confirms Secretary of State, Razvan Murgescu's affirmation that 'Branding Bucharest implies not only the promotion of the city but also the solving of its problems beforehand '. Based on current research findings we could conclude that the final outcome will portray a city brand strategy that will convey a common vision through both an internal and external message, based on the unification of all the stakeholders in partnerships and relational networks.

> ## "Don't be afraid to borrow from the best, while trying it your own way." Keith Dinnie

Iulia Gramon-Suba
Design and Branding Strategy

Internal Branding within UK SME Retail Operations

Enhancing SME retailers by embracing an inward-facing, design-led approach

More than a third of consumer spending goes through 286,680 retail outlets in the UK. Retail employs 11% of the total UK workforce and the number of non-food independent retailers grew 5.6% in 2009. The industry is resilient, but the low-growth environment has led to substantial challenges to retailers as fewer shopping trips make the significance of every shopper interaction more important. Consumers are becoming ever more complex and multi-faceted whilst retailers struggle with the integration of multi-channel communication to provide a seamless proposition for the customer. In addition, due to the recession: lowered budgets are demanding high ROI and staff morale is often stated as being at an all-time low. Highly visible well branded major franchises have remained buoyant. But is this true for smaller operations? Research shows that in smaller retail franchising guidelines, the regular absence of words 'brand' or 'design' is very telling. This initial research supports the Design Council's findings which state SMEs fall behind larger enterprises in seeing the value of design as an integral part of their business. Much of the design implemented tends to be focused outwards, solely on connecting with customers. This outward view may be flawed as internal branding and traditional external branding should and can, complement rather than compete. Retailers have the opportunity to create a human-centred brand at relatively little outlay to look inwards and engage with their own staff resulting in improved on-brand articulation and delivery to customers. Internal branding has many benefits such as increased staff motivation, energy and pride. The challenge is how to ensure SMEs understand the value of building and implementing a strong internal brand in times of austerity. SMEs have the ability to move quickly and innovate to differentiate from bigger rivals so are best placed to take advantage of opportunities in the future. This research investigates and discusses the role of design within internal branding, examines current practices and will pragmatically discover opportunities for its implementation within SME retail operations. Franchised operations are of particular interest due to the core need for a replicatable concept and system. Emerging research tools are used to understand the relationship of internal branding from the company employee perspective (Cultural Probe) and Company Management (Contextual Enquiry) with the outcome of the production of strategic imperatives to create a strong internal brand capable of inspiring and engaging staff whilst integrating multiple components such as customer interactions, employee communications and corporate philosophy.

Kylie Arthur
Design and Brand Strategy

Delivering Compelling Brands through Experience Design
Translating brand values into an interactive experience in the retail industry

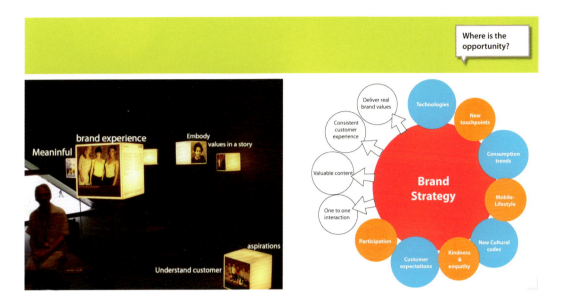

"Consumers today connect with brands in fundamentally new ways. That means traditional strategies must be redesigned to accord with how brand relationships have changed." The consumer decision journey has changed towards people sharing, rather than receiving experiences. As a consequence collaborative consumption tends to play a major role in the customer decision journey. There is also a major shift from passive behavior to a participation culture. All of these facts raise the customer consciousness and expectations demanding more meaningful and engaging experiences. According to Warren Hutchinson in "Branding Multichannel Experiences", brand experience could be defined as, "Optimized Usability + Brand behaviour = Brand experience". Brand strategies should be evolved into interactive relationships translating brand values into tangible actions, creating an improved compelling experience, where brands adapt to customer expectations increasing

customer loyalty and raising brand awareness. "Satisfaction and meaning are established only by experiences that customers have, across each of the brand's touch points over the course of the time, these customer experiences define what the brand is all about." Designing meaningful brand experiences could empower the the identity of the brand by delivering values through developing a transparent and consistent attachment between customer expectations and the brand profile. The design research will be carried out with a combination primary and secondary research methods aimed at developing future brand scenarios and forecasting responses and demands of emerging trends in cultural, technological and social consumption areas. Stakeholders from design, branding and the retail industry might concentrate on how design is fundamental when talking about branding futures. As designers we need to translate brand value into a reliable customer journey. Furthermore the retail environment must then become more of a brand experience opportunity, aiming to develop relationships with customers, creating scenarios that allow customers to feel the brand experience, projecting emotions and expectations rather than focusing on selling goods.

"Develop consumer-driven brand values and associated design criteria for an ideal experience and deliver brand meaning through design" C.Rockwell

Maria Paula Martinez Rodriguez
Design and Branding Strategy

Shifting the Public Perception of Used Car Dealerships

Building trust through multi-channel experience design and branding strategy

Design, apart from its obvious creative characteristics, is now acknowledged as a problem solving process. During research on the recent meta-recession consumer behaviour and, in particular its effect on the automotive industry used car market, two key issues emerged as clearly identifiable problems:

1. According to research into 700 consumers by Euro RSCG Biss Lancaster, only 18 percent of consumers trust what they read, see or hear in the media, with only 11% trusting the experts.

2. As reported by Consumer Direct, there are concerns that consumer satisfaction and trust in this market remains low. Consumer Direct continues to receive a flow of complaints from consumers, setting a perception that certain players in the market are seen as untrustworthy. Identifying this problem led to defining the aim for the project: 'To develop a design and branding strategy for used car dealers to build trust and reliability.' The methodology being used requires critical research methods to identifiy opportunities for design by finding ways to enhance the experience of customers through multi-channel experience design and a consistent branding strategy.

In a consumer's mind the past experience of a brand is significantly linked to the perception of this brand. For consumers the value of each brand is conveyed through positive product experience and lasting brand loyalty.

In the suffering market of used car dealerships, trust is the missing ingredient to build customer loyalty. Ted Minnini sees trust as a composite of four main components: 1. Connectivity, 2. Innovation and Service, 3. Fair price and 4. Transparency, Honesty and Authenticity.

If the experience of the buyers is enhanced and an honest and authentic relationship is developed between customers and brands, brand loyalty and advocacy is achieved. According to the law of Fellowship the whole market could take advantage of the positive publicity between consumers, that will end up shifting to a positive perception of the used car dealership.

> "Consumers are hungry for values they can put their trust in." Ted Mininni

Theodore Tsikolis

Design and Branding Strategy

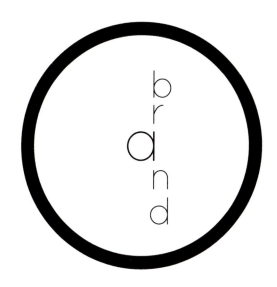

CLIENT: a brand

Now add
colourings
flavourings
& community

A particularly contemporary problem that brands suffer is that their owners seem frequently to lose control of them. What causes this?

I have been somewhat unsatisfied with the brands that have been designed. Not because they were badly designed but because they felt artificial. It took some time before I realised that my work was based on the assumption that brands must necessarily be artificial constructs. I have since questioned that assumption. After all, it is unsatisfying to think this way when brands themselves aspire to such genuine expressions of authenticity.

These artificial constructs have become increasingly sophisticated. Brands now have personalities, generate emotional responses and even connect spiritually. However,we should not lose sight of the fact that brands are not real.

Today consumers have more to say about brands partly because they have noticed this lack of authenticity. Faith Popcorn cites brand promiscuity to be a top ten trend for 2011. Wally Olins says the customer is now firmly in charge. Clearly the consumer has become increasingly active and brand owners are increasingly losing control over their brands.

A likely future scenario appears to be much closer to brand through dialogue. Furthermore, this dialogue can be expressed by brand stakeholders not just in brand promiscuity or verbal commentary, but in the growing desire for active involvement. This desire is also evidenced in the trend amongst today's consumers for collecting experiences rather than just passively consuming products or services.

A way to explore these ideas of brand as dialogue is to look into brands that have community at their centre. Social enterprise brands offer some of the most promising forms of meaningful involvement for multiple stakeholders since they frequently adopt forms of participation from the community to operate. Unravelling the connection between brand and the participation of its stakeholders should lead to useful guidelines for creating authentic brand experiences. By tapping into the motivations behind participation, this can help to strengthen and sustain social enterprise brands that are not artificial constructs, but expressions of the dialogue between real participants.

This research will practice what it preaches and mix contextual enquiry with objective insights from the author's active participation in creating and sustaining social enterprise brands.

"We don't own or manage the brand [The People's Supermarket], the community does." Rachael. E. Baird, TiltStudio, USA

Matthew Watts
Design and Branding Strategy

The Perception of Colour in Packaging Design

Psychological associations people have with colour: a style guide for brands

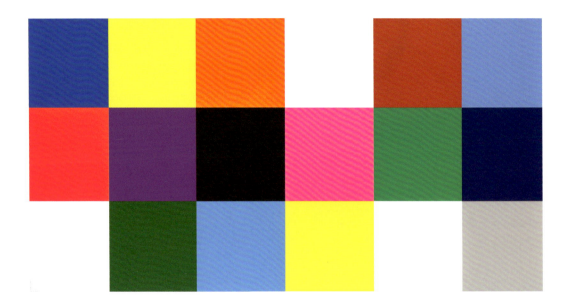

As you go to the store to do your weekly shopping, you may notice how the shelves in each aisle are packed with hundreds of different products and brands. Each year, the number of different products increases and the available shelf space for each one gets reduced to make room for more variety. In fact, your eye rests on each package for about 0.03 seconds as you pass down the store aisle. This has increased the competition for shelf space in order to increase their chances of you seeing their product. This makes the packaging even more vital to the success of a product as "80 percent of a consumer's decision is now made at the shelf, with an emphasis on first impressions and price than on brand reputation and purchasing history". Packaging is often the first impression that a consumer gets of a particular product, so it is important that care is taken in its design. Colour can play a key role in developing the right balance between images and text. Colours all have an inherent meaning in them and they can be used to a brand's advantage when designing a package. By taking advantage of these subconscious associations,

> "Colour must immediately attract the consumer's eye, convey the message of what the product is all about...and help to make the sale." Leatrice Eiseman

a brand can convey a more complete message of what their product is without overcrowding the design. For example, pink is often thought of as sweet, youthful, or happy and therefore would work well on packaging for sweet items to subconsciously persuade the consumer that their brand is sweeter than another. Burgundy and dark brown are perceived as rich or luxurious and would work well for a more expensive brand of chocolate. Consumers would feel as if they were indulging themselves in a more luxurious treat because of the chosen packaging colour. Through colour and package studies, the research aims to delve into subconscious associations with colour in order to develop a packaging style guide that brands could use to strategically choose their product's colour palettes. This would help brands create more unique, and in some ways more efficient, packaging that would increase their visibility on supermarket shelves. It would also help to convey brand values in a subtle way that would entice more consumers to purchase their product.

Ashley Brooks
Design and Branding Strategy

Three In One Headphones
Three kinds of headphones, one set

Balance Emotion
Using colour to influence emotions

Today, consumerism is increasing at an exponential rate; companies are taking strategic approaches bombarding consumers with a paradigm influencing attitudes towards products and their use. This encourages a buying frenzy, where replacing existing products before they become dated becomes a priority in the minds of modern day consumers. This type of behaviour has negative effects on the environment because of an increase in waste production; which leads to ecological depletion and causes a drain on new material resources needed for the manufacturing of the new products demanded. A new sustainable attitude is needed in consumers focusing on preservation and longevity encoded in the DNA of new products. This change will encourage people to keep their products longer, with these products easily repairable, discouraging throw-away culture. The 'three-in-one headphone set' is a product encouraging this philosophical approach, promoting new ways of thinking. Headphone products are being used more often in people's everyday activities, in turn creating a new lifestyle product. This is where the opportunity for the project lies, as different activities require different types of headphone compensating for the difference in activity the user might participate in. This will encourage the sustainable product philosophy as the user will not need to buy different headphones as they will have an all in one unit.

The core of the project revolves around research showing that people's moods can be influenced by elements of colour in their environment.

The 'Tree House' is a flexible furniture system, which creates seating and table surfaces in a variety of configurations. The design uses an embedded colour changing system, which senses people's biological data and adapts the colour rendition of a space to trigger a change in mood. For example, if a person was sensed to be feeling unhappy, the colour would change to a bright and active colour such as orange.

Tree House is made up of multiples of a single component, which can also create temporary private space. The Tree House can be used in both private and in public environments such as libraries, coffee shops and shopping malls. Multiple installations of the Tree House in an environment will produce a spectrum of coloured effects responding intuitively to the emotions of the users of the space and ultimately affecting positive moods.

The form of the 'Tree House' is inspired from the organic, yet structured intricacies found in nature. Dieter Rams' principles of good design were followed to make the design sustainable and minimalist.

Andres Cervantes
Integrated Product Design

Hongchen Zhang

A Unique Experience through Visual Merchandising

Generating a strategy to strengthen brand image through store design

"It is your imagination that needs to be stimulated. Once that happens, the rest is easy. The merchandise is always the leader."

Today, one of the most important points of branding is visual merchandising. Companies can directly express themselves to their customers with their shop environments. The purpose of visual merchandising is to show the product properly, leading customers to the product, and creating a convenient environment which can provoke the senses of the customer to make products purchasable.

Furthermore, companies can augment, reinforce or defuse their brand images through the right merchandising strategy. Today's customers want to be included in this branding journey. Thus, designing consumer experiences at multiple points of interaction is the essential rule of enhancing brand image.

Heavy competition has made brands focus on their store design because they realised that building a strong connection with customers is an important component to profitability. Therefore, good store design is crucial to show the best brand image.

In light of these valuable insights, this research is being carried out in terms of the relationship between brand image and the store design. The market leader of Turkey's home appliances industry, Arçelik has been picked to generate a strategy that will align the visual merchandising design to the overall brand image. As there is a perceived gap between Arçelik's innovative and energetic brand image and the way it is promoted in its exclusive shops.

At the end of this project, it is hoped that both Arçelik, the owners of Arçelik's Vendors and the customers will benefit in terms of a strong brand image, customer satisfaction and profitability.

In terms of primary sources for this research, interviews have been conducted with Marketing and Sales Directors, Retail Support Manager and Merge Team and Sale Specialists of Arçelik. Digital ethnographic observation was carried out in Arçelik's exclusive shops with professional visual merchandising team.

Concurrently Customer Surveys have been conducted since the beginning of April and final negotiations with directors of Arçelik will be planned after customer survey analysis and passive observation.

Finally, this research will generally give valuable insights about the importance of visual merchandising to strengthen brand image in terms of hard or soft trends. Arçelik can gain important clues to understanding their current weaknesses and seeing their customers' response to an enhanced brand image.

Fatos Ceren Tan

Design and Branding Strategy

In the UK, 99.9% of businesses are small to medium sized enterprises. Differentiation can be challenging. The recession is still negatively affecting SMEs. Typically, organisations that value design tend to outperform those that do not. I am fascinated with how guerrilla design is effective for large corporations. The focus of this project is to develop a strategy for identifying efficient, imaginative ways to promote and differentiate SME businesses through innovative guerrilla design. For branding, marketing and advertising activities this involves using unconventional methods of getting messages across to consumers. Due to the emotional and or interactive nature of guerrilla design, SMEs can enhance the two-way communication. What ties guerrilla design to advertising, branding and marketing methods is the fact it can be cheap, quick and easy to implement. For struggling SMEs, expensive advertising campaigns are not an option, creating a catch 22. Little money, few ads: few ads, little business.

Jay Conrad Levinson is a key figure in guerrilla design and marketing. He coined the term in 1984 due to the similarity of the unconventional guerrilla warfare methods, and today it is still a fairly new paradigm. His books describe tactics to leverage guerrilla design in order to gain publicity. By performing action research and liaising with SMEs in the West London area, a picture been built up of the pressures of today's business and different views on guerrilla design.

During a presentation at Brunel University, Simon Black, Group Strategy Partner at Design Bridge stated "good interpretation is great, originality is not vital, but test your lateral-thinking, move it on". Black was speaking about branding and the 'thinking outside the box', which is extremely important for innovation and growth. In terms of guerrilla design for branding, this relates to unconventionality as a key aspect for publicity and promotion.

Based on the research to date, the final outcome of the project should be able to prove that guerrilla design can be leveraged by SMEs, whether design-led organisations or not. At the end of the project, a simple, prospective strategy that can be adapted from one business to another, encouraging creativity to create successful guerrilla design approaches for advertising, branding and marketing will have been developed.

> "Good interpretation is great, originality is not vital, but test your lateral-thinking, move it on." Simon Black

Claire Hall
Design and Branding Strategy

Inspired by Nature

Using biomimicry to enhance the creative process for product designers

Biomimicry is a design principle that seeks information from nature to solve human problems. Traditional design methods rely on human innovation, whereas nature has limitless possibilities of inspiration. Using nature as inspiration is what inspired the focus of this research project. The aim of this research is "To develop a tool based on the principles of biomimicry that product designers can use for increasing and broadening concept generation by extracting principles of biology and applying these phenomena to design problems."

To gain more specific knowledge relevant to the aim, a large collection of interviews from academics, designers, biologists, biomimetic enthusiasts and people who don't have experience with biomimicry will ensure a well rounded body of opinions and experiences. These opinions can be useful to help understand the ideation stage and explore how nature can be incorporated to help enhance this process. By synthesising the research a "tool" can be

developed to benefit product designers.

The Biomimicry Guild is a nature inspired consultancy specializing in teaching and helping companies learn from nature. From contacting the Biomimicry Guild details about the services and tools they offer to help clients learn about biomimicry and how to apply it to their products or services have been acquired. Biologists at the Design Table or BaDT's, are people that serve as communicators between the business and natural worlds. One BaDT, Tim McGee, commented via email on how biomimicry can be used as a way to generate creative ideas saying, "Part of the practice and art of biomimicry is engaging with multiple fields in creative ways around a common universal language (nature)". Nature can help in escaping the "norm," or "expected" and lead designers into having a new perspective and broader range of ideas and solutions. Carlo Santulli, a materials scientist, agrees that, "a parallel investigation into nature throughout the design process will lead to a greater range of ideas".

This research plans to show that nature is a valuable asset that can help product designers during the brainstorming process; and might provoke an idea that never would have been thought of without a conscious exploration into the natural world around us.

"A parallel investigation into nature throughout the design process will lead to a greater range of ideas" Santulli, 2011

Heather Bybee

Design Strategy and Innovation

©Polevoi, Ethnographics

We are living through a financial crisis in a multi-cultural society; with high living costs, poverty, low educational outcomes and anti-social behaviour. Soaring childcare costs are preventing parents from going to work.

Children's experiences and development during their early years influence their basic learning, educational attainment, economic participation and health. There is, however, more emphasis (through Ofsted and the Early Years Foundation Stage in the UK) on academic development in early education than in emotional development – which is considered a strong foundation for preparing children for school and life.

The Department for Education's recent review on the Early Years education programme shows that social and emotional aspects of education should be emphasised. A lack of emotional development in early education can lead to anti-social behaviour and poor communication skills. This must be addressed in order to prepare children for a multi-cultural environment in an industrialised society.

Antidote opines: "Our school system was designed for a society that was relationship rich and information poor. To thrive in the very different circumstances of today, we need to develop schools that will nurture people's capacity to connect with each other".

Children would benefit from socio-emotional development from role models both at home and in early education. Childcare staff could build a strong relationship with both parents and the community to help increase inclusivity and support in the children's environment. Parents would have the oppotunity to participate within their child's education with the support of the healthcare provider. The community would have the ability to donate or be voluntarily engaged in local early years services, in activities such as sports, drama or arts and crafts. The government would benefit from greater community and local support which would put less pressure on public services. The return on public investment in high quality early years education can include a reduction of social problems and inequality and an increase in productivity and GDP growth.

Design research methods such as ethnographic research, interviews and creative focus groups will allow for great insight into the early education programmes and collaboration from the most relevant people in this industry.

Potential outcomes for this project include the changing perspectives of emotional literacy, children having the increased ability to manage and express emotions in a social context, environments and activities that promote the development of socio-emotional skills in children; and the increased involvement of both parents and the community providing support and creating a strong community bond.

Luiza Frederico
Design Strategy and Innovation

dt98cai

What Good is Brunel?

The 2011 Made in Brunel Show and this book are most definitely about this year's crop of talented graduates and I have no wish to distract attention from their degree showcase. Rather I offer this single point of reference for how a Brunel design degree enabled this writer to follow a passion for design as a career, as I trust it will do for the Class of 2011…

To put things in context, I graduated from the Industrial Design and Technology course at Brunel in 2002.

When we graduated, the talk of my year group was that the Runnymede, aka 'Design', campus was due to close and the courses relocate to the Uxbridge campus. The biggest question in our minds was how would this impact Design at Brunel, all it stood for and all we had given it? Having had the opportunity of a sneak preview of the 2011 Made in Brunel team, I am both relieved and proud to see that the university where my year gave four hard years has maintained the highest level of international design education and the calibre of the graduating students is still second to none.

Brunel equips every student with a veritable arsenal of skills and design knowledge that is crying out to be applied and to be developed in a commercial environment. Unfortunately it takes time for graduates to be able to look back and truly appreciate the incredible basis for a professional design career that Brunel Design gave them.

Much is said about the breadth of subjects and skills that are covered by the design courses at Brunel. Some employers who have not had first person experiences of the university are often a little bewildered by the person before them at interview. With such a wide skill base it might at first be hard to see how this person can be integrated into an existing design team structure. Those in the know already understand that this is the very key point of a Brunel graduate: 'knowledge and adaptability'.

We stick to what we know and therefore it was no accident that my first design manager was himself an Old Brunelian. In hiring a Brunel graduate, he knew exactly what he was getting: a versatile, creative, team member who was commercially viable direct from University. Today this is the same reason that I count Brunel graduates amongst my company's interns and in others' programmes and where they have also been recommended to look to Brunel graduates.

One thing that is quite clear, reflecting on my own career development and those of my peers, is that with the adaptability and versatility inherent to Brunel graduates, the design courses at the university excel at creating design leaders, managers and innovators. Brunel really builds designers to lead.

For some this can be a quicker journey than for others… Having had regular business with the Far East over the first two years of my employment, I chose to move to China full time in late 2005. In my first role I referred to every part of my Brunel education and guidance under my Old Brunelian manager to set up an industrial design department within a manufacturing facility in China.

Fraught with cultural differences and having to pilot the education of the company in how design could be an asset to more than styling alone, the first year was an amazing challenge and opportunity.

Over the next three years my role within the company developed to include full creative direction over industrial design, packaging solutions and marketing materials. Again, it was the adaptability that the university course promoted that enabled me to step into these positions.

The Brunel course encouraged us as students to be creative and exercise our passion. It did not restrict us, but equipped us with an array of skills and tools that we could later rely on to adapt and use as necessary in the business world.
I do not consider my design education complete. Being in China by itself provides me with daily challenges in helping develop the local design industry with the knowledge and skill that my Brunel education and career experiences have taught me. Now having added a leading Chinese industrial design consultancy and a Chinese user interaction design firm to my employment record I have managed to bridge some of the gap between East and West and learnt more.

The domestic industrial design industry in China is expanding at a furious pace yet still remains somewhat in its infancy in developing a design history and a shared experience bank to refer back to. A design process we would understand in the West is taught at universities here, but as a relatively new industry, companies and clients that are using design services generally do not have the wider understanding that industrial design provides more than purely stylistic value alone.

China has been in a rapid development phase since the Cultural Revolution and shows little sign of slowing down. It is easy therefore to understand why the local design industry lacks a larger, culturally relevant, direction for design. As a mirror to the rapid rise of the country, Chinese design sells 'new'. Unfortunately 'new' becomes 'old' very quickly. This promotes a constant turnover of new products and means that where brand value is high, brand experience is not. If you are constantly chasing 'new' the quality of 'now' is likely to suffer.

As more Western products and brands move to China, things are slowly starting to change. More coherent brand values, language and experiences are becoming available for general consumption. In this environment, Chinese brands will have to adopt similar strategies in order to remain desirable and competitive.

It is therefore within this 'design culture' and understanding of both East and West design industries, that my current career phase sees me taking the next step. I am putting the 'Brunel builds designers to lead' principle into complete practice as I start my own design company in Hangzhou, China. The intention to match the best of East and West will enable the company to fill a growing niche in China that sees 'good' design having more weight than 'new' alone.

The future is definitely set to be full of personal challenges, but these are the challenges that excite and without them my passion for design would remain unfulfilled. My hope is that the 2011 group of Brunel design graduates can all realise similar challenges and opportunities. I wish them all every success in their chosen careers - a Brunel design degree is just the start.

dt98cai, Founder, CAIS Design
Graduated in 2002

"Brunel really builds designers to lead"

Future Urban Homes
Designing for the 2050 population

By 2050, 75% of the world's population will live in cities. Because of the influx of people in city living, a greater focus on designing for an urban population is necessary to maintain a good standard of living. Emerging demographic and social changes, combined with a push toward more sustainable living and rapid advancements in technology all inspired the focus of exploring the role of design in the future of urban living. To explore this topic, research was conducted into current and future trends in demographics, sustainability and technology by reading trend forecasting websites and consulting expert opinion. To analyse the research a timeline of events was created to see how they interact with each other. From the timeline, some events were highlighted that could have a profound effect on the role of design in the future. Some events include the possibility of the retirement age extending to seventy years, oil demand outpacing supply, global food and water crises, and human-like A.I. becoming reality.

These researched predictions were used to form a picture of what urban living may look like in 2050. These ideas fall into three categories: the structure of the home, the products in the home, and the people who inhabit them. In the future, flexible and modular living spaces will allow for a more efficient use of space and accommodate a growing population. Alison Brooks of Alison Brook Architects (ABA), believes it is a designer's "duty to improve the quality of life for urban dwellers, beginning with housing." Buildings can also be designed to collect rainwater or incorporate indoor gardens to help supplement food supplies. Intelligent products that are interactive and responsive to your actions can enable humans to overcome future environmental and social obstacles. The implementation of 3D printers in the home environment will transform shopping experiences into completely instantaneous and customised experiences. People who inhabit homes of the future will be more virtually connected with the world around them and may include holographic capabilities realistic enough to enable real time interaction with others.

From trend research and consulting with experts it is apparent that design's role is always increasing. It will be necessary for designers in the future to be more holistic innovators of life because the boundary between humans and products will 'blur'. Zaha Hadid, of Zaha Hadid Architects, believes that we are "restructuring away from an industrial mass society towards a society with much greater degrees of complexity and dynamism in people's lives". Therefore, a holistic design approach is vital to handling future cultural complexity to maintain a sustainable and high standard quality of life.

H. Bybee, B. Kirk, S. Lee, M. Kim, L. Frederico, H. Lee
Design Strategy and Innovation

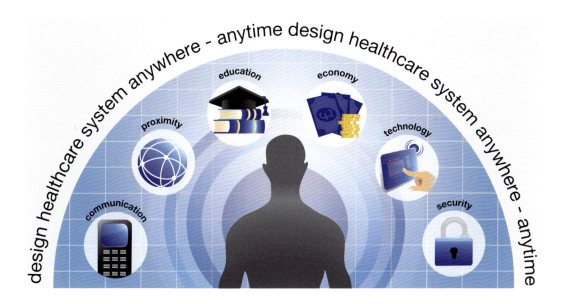

This project focused on defining problems and opportunities for design in different national healthcare systems, focusing on patients rather than on technology, a human-centred approach. The methodology included hard data collection, experts' opinion and Delphi. Historical research helped us in combination with current research; through extrapolation of the research we came up with our future scenarios.

Separate literature reviews were used to gather background information for the research which, when conbined, formed the basis for the main framework and timeline. The timeline consisted of four milestones; 2011, 2030, 2050, and 2080. At present the role of design is rather technologically oriented and the problems of high cost, inconvenient geographical position, and lack of staff are hard to overcome.

In 2030, it is thought that design will have the role of an information manager and help transform the system to be more reliable and balanced. During 2050, design will facilitate the collaboration of patient and doctor through commoditisation of services and high technology. By 2080, the form of the healthcare system will have changed completely, with the role of design adapting to 'modern' needs. Emotional and psychological health will be prioritised since healthcare will be accessible to anyone, anywhere, at anytime through self administration of healthcare.

To conclude, in the future, design will have to adapt to the changing needs of the patients and tackle challenges more efficiently, cost effectively, locally and securely. Moreover, there will be a shift from product-centred design to patient-and-system oriented design.

It is clear that the beneficiaries from this research will be medical organisations that wish to conduct research into the future of healthcare, medical practitioners and patients, and governments that might decide to invest in the healthcare system.

T. Tsikolis, J. Zambrano Novoa, H. Shim, D. Tzarela, K. Trigkaki, E. Song
Design and Branding Strategy

Future Breakfast Routines in China

Reintroducing healthy eating routines to workers in China

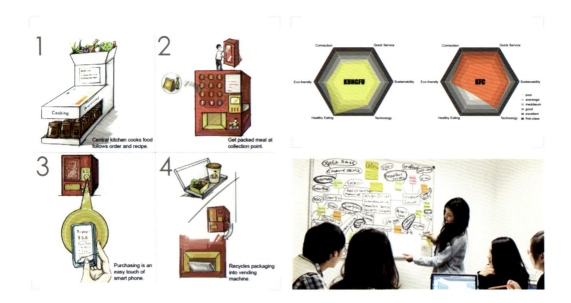

Research shows that 60% of office workers have only 5-10 minutes to eat breakfast and 26% do not have breakfast in major cities in China.

To understand the issue, design future forecasting and design thinking are paramount to help turn extreme complexities into legitimate opportunities for the future.

To begin, design thinking was applied to a PEST analysis, the results showed that many Chinese breakfast fast-food restaurants lack sufficient design inputs to link technologies and businesses. Both hard data and soft trends around the issue were collected; the international and local fast-food brands market share in China showed that KFC holds 36% while Kungfu only has 11%. Food habits have seen enormous permanence over the past 100 years in China, so how can Chinese brands like Kungfu, which integrate healthy Chinese-food tradition and Western fast-food service into business, gain more market share? They may be the solution to improve Chinese office workers' breakfast routine in the future.

Delphi Forecasting was used with a panel of eight experts in order to capture the key words about future developments in fast-food restaurants, which are eco-friendly, technologically connected, healthful, sustainable and quick serve. Based on research and group workshops, application of The Competitive Analysis of Chinese and international fast-food brands helped to visually and theoretically forecast the possibilities of design for the future of Kungfu. The result showed that Kungfu is highly recognised by Chinese consumers as a brand that provides healthy Chinese food but, compared with its biggest competitor, KFC, it lacks sufficient application of design to build a convenient environment for consumption.

The project outcome displays the future environment for consumption of a Kungfu Client. A central kitchen cooks food following the customer's order. The packed meal is then delivered to a collection point. The client collects his meal in a vending machine with his consumption data in the system. The purchasing activity will be as easy as touching a smart phone. After enjoying the meal, the client recycles the packaging into the vending machine.

In the future, the application of design will provide Chinese fast food brands with a highly efficient operating system which ensures both high quality and quick customer service. Wally Olins says Chinese brands should believe in that being an original Chinese brand is good enough. Chinese brand owners should take advantage of employing design-driven innovation to build strong brands.

X. Yu, H. Dong, Q. Xiang, S. Yang, W. Ye, Q. Ge
Design and Branding Strategy

Increasing emphasis on the importance of change within the educational system coupled with user experience of education over two decades resulted in a desire to find areas in which design impacts upon the educational landscape. According to Robinson almost every education system in the world is being reformed. Creativity and collaboration should be a key principle for this changing process.

By reviewing the history of educational systems a linear process was identified, with the beneficiaries holding a passive role. Recent technological changes have made information more widely reachable: creating the conditions for accessibility and multidisciplinary integration to reshape the way education is being delivered.

A PEST analysis identified the current and emerging micro and macro trends in education, allowing for the analysis of the different areas where design could have greatest impact assisting educational change. Three areas identified were: Process, Technology and Environment.

Design thinking was integrated into the Process area, with collaboration and multidisciplinary work as groundbreaking tools. Educational 'fablabs' would be created, and learning-by-doing would become the norm. Sharing the learning process with peers would result in users with broad, but not necessarily deep, quality knowledge.

Professor Heinz Wolff mentions in an interview, "we need to educate in a more collaborative process that enables versatility in a person."

In the second scenario (2030), technological developments lead to an individualised approach to the education process, having beneficiaries customising education to their needs and interests and creating a network of exchangeable knowledge. With the aid of digital platforms such as holograms, individuals would be able to barter knowledge with their peers. In this scenario, individuals benefit from obtaining information in areas they are not experts in. The main outcome would be highly specialised and skilled individuals.

The third scenario (2050) draws the convergence point between Process, Technology and Environment. Using the right set of technological tools and having the right process embedded into those tools, the education process can be held in any possible physical space. Knowledge will be a common asset, available for everyone through a virtual Internet cloud.

To conclude, in the future Design will transform education by facilitating a multidisciplinary collaboration amongst experts in different fields of knowledge, in order to create ideal Environments, Technologies and Processes for the education system.

C. Hall, K. Arthur, C. Rojas Monserratte, N. Al-Twal, G. Gana, A. Pearse
Design and Branding Strategy

The Future of Branding

Considering the importance of branding in thirty years time

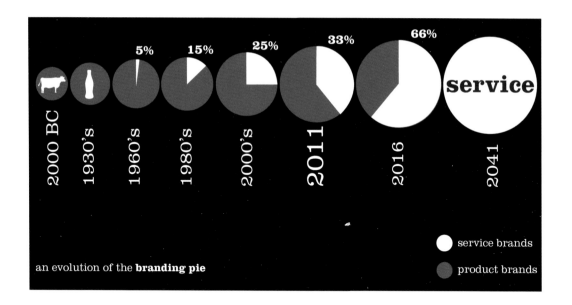

an evolution of the **branding pie**

Brands cannot rely on the 'ignorance' of their consumers anymore. Consumers are more aware than ever before and they want to believe that they are buying into something that they themselves believe in or want to be associated with. Brands have to spend more time and resources listening and responding to their customers. The brands that achieve this will be the ones that will flourish. The research was split into two stages. In stage one six current essential elements for successful branding were identified: sustainability, communities, co-creation, story-telling, transparency and privacy, and design as interpreter. In stage two, the primary research consisted of interviews, lectures and discussions on social networking sites in order to discuss theories with brand, marketing and design professionals worldwide, gathering opinions from over 35 individuals.

Based on this research, the vision of the future of branding relies on a new concept developed called 'Reversed Co-creation.' It refers to consumers creating the brand through demand (unlike co-creation, where organisations ask for consumers' input). As a result, design's role as an interpreter will grow. As the consumers become more influential, design will translate their demands into tangible services, products and brands. There is some evidence of this trend today through consumer-created online brands, but it is believed that, in the future, this trend will become more widespread. As Jay Olson, Brand and Marketing Strategist at QuadCreative, says, "consumer engagement as a co-creator in the development of a brand's equities is key."

Based on this concept, two future scenarios have been developed. In five years time the prediction is that service will become the main differentiator, the factor that will be most important in building consumer/brand connections. This trend can be seen to be taking shape today by looking at brands that have achieved success through the incorporation of complementary services, a strategy that is setting them apart from their competitors.

The second prediction: in thirty years' time fast-moving consumer good (FMCG) brands may become unnecessary. As 3D printing and replicator technologies advance, FMCG brands may become redundant and therefore services will remain to rule. In a recent lecture Warren Hutchinson, Founder and Creative Director at SomeOneElse, emphasised this point stating: "Brands must learn how to interact and walk the talk."

In the future design will be the most important connection between consumer and brand, allowing for clear two-way communication, a point highlighted by Simon Tempest, Senior VP Marketing at ACCO Brands; "brands are a tired marketing tool, becoming loved (really valued) by the consumer is where the future is."

J. McCormack, A. Oprisan, A. Brooks, I. Gramon-Suba, J. Sook Han, Y. Li

Design and Branding Strategy

LED Light Ioniser touch screen
Nature + Light therapy

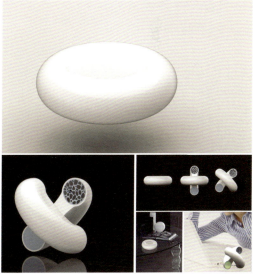

LG Electronics Design Europe came to Brunel postgraduate courses of Design Strategy and Innovation, Brand Strategy, and Integrated Product Development to seek external ideas for two new products: an air ioniser and LED light. The challenge was to create innovative concepts with emphasis placed on the user experience to build a scenario, usage scene, functions, and unique design forms to express features.

Students initially submitted a written proposal of an initial idea for one of the two products. Students were then chosen to work for eight weeks on their own product idea. A workshop at the LG Design Europe office was a chance to learn about the company, gain more specifics about the project, and brainstorm ideas with others. Final presentations showcased a wide variety of ideas for air ionisers and LED lighting solutions.

H. Park, H. Bybee, K. Arthur, M. Suh, S. Woo
Brunel Design Masters

Design Bridge and KFC
Student Innovation Challenge 2011

The international brand design agency Design Bridge partnered with Brunel Design Masters courses to propose the challenge of redesigning an element of the KFC customer journey. Students representing Design Strategy and Innovation, Brand Strategy, and Integrated Product Design made up teams to work together on the project for four short and intense weeks. Each team took part in experiencing the KFC customer journey and critically analysed it to look for opportunities for improvement. Groups conducted interviews with employees, observed customers, and researched the KFC brand. Final ideas were shown in sketches, renderings, and 3D mock up models. Each group's final design was presented to KFC and Design Bridge and displayed at the Packaging Innovations exhibition at the NEC in Birmingham, England.

H. Bybee, K. Arthur, C. Hall, N. Horan, M. Suh, M. Martinez Rodriguez, G. Chung, I. Yang, I. Gramon-Suba, T. Tsikolis, A. Brooks, N. Amjadi

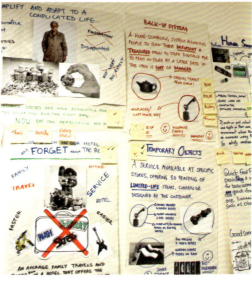

Since 2008, the École Nationale Supérieure des Mines de Saint Étienne partnered with Brunel University's graduate level design courses to offer a one week multidisciplinary workshop. Students across all engineering subjects, design, arts, and students from Design Strategy and Branding from Brunel gather to explore a specific topic area to experience a true 'co-creation' process. This year the project explored the future possibilities of "personal fabrication systems." Starting with a large group brainstorm, ideas were then focused in small group settings, then expanded out into larger and more detailed presentations. Concepts from this year included: using personal fabrication for travel, medicine, toys, or fashion. Projects produced physical mock-ups, designed brands or formulated business models.

Designs and strategies which aim to reduce environmental impact.

humanistic innovation   19

sustainable innovation   197

technical innovation   245

digital innovation   321

network directory   353

Future Concepts for BIC?

BIC has been a global brand for over 60 years, producing products people trust and use every day. Unfortunately the majority of BIC products are made of plastic that cannot be recycled. Increasingly, companies are faced with environmental and social pressures to improve the impact that they have on the planet. Subsequently, BIC must look to adapt their strategy in the coming years. As part of a "More Responsible Future", BIC would look to improve their products both socially and environmentally. Each product would have a positive effect on the user, addressing an element of BIC's responsibility. With an increasingly responsible attitude to future development, it is hoped BIC will continue to positively strengthen its brand image.

Seed Storage and Planter
Future concept for BIC

More people than ever before are moving into cities and consequently the pace of life is increasing. We are settling for fast packaged food and an indoor lifestyle, neglecting the small green spaces available. As children grow up in this culture they are becoming increasingly unaware of where their food is coming from and how it develops. As a part of BIC's more responsible future this product encourages people to realise the joy of planting, growing, and harvesting their own fresh vegetables. This will in turn promote the use of plant pots, small gardens and allotments. These refillable seed planters allow the user to store, take pride in and enjoy the plants lifecycle. The simple planting mechanism means that in just one click the seed can be planted at the correct depth, resulting in the best possible harvest.

Rohin Odell
Product Design Engineering

BIC are currently perceived as a 'disposable company'. They produce a large range of plastic products with a short lifespan with little or no recycling potential. This has prompted the need for BIC's 'more responsible future'. The BIC Paper Maker is a product that encourages recycling within the home by allowing the user to be a part of the recycling process. The process of making paper is relatively simple; it involves using your waste paper to create a pulp, which is poured into the BIC product. A mesh frame is then lowered into the pulp before it is emptied and dried. The frame can then be reinserted until the pulp has run out, creating multiple sheets of personalised paper. The product can be used by children and the aim is to help make recycling fun, whilst increasing awareness about recycling.

Matt Parrish
Product Design

Ink Cartridge Refill Station
Future concept for BIC

BIC products have always had a negative reputation when it comes to disposal. A new direction would see BIC turning this on its head so that their products help to recycle and re-use existing plastic products, instead of being instantly disposable. This product refills printer ink cartridges. This allows the user to keep and reuse their plastic cartridges time and time again. This would also save the ink that gets left in cartridges after each use. Through the cartridge refill station only bulk purchases of ink would need to be made, saving on numerous plastic cartridges throughout the year and reducing consumer plastic consumption.

Fred Swallow
Industrial Design and Technology

BIC's products are known for their reliability and disposability, but in recent times the human impact on the environment is becoming more and more globally important. Many of BIC's products end up at landfill sites due to their plastic nature and not being recyclable. As part of a new brand direction for BIC "A More Responsible Future", this concept is a hand-held cutter that removes the heads from BIC's cheap disposable razors (containing metal blades), thus enabling the plastic handles to be recycled. The cutter also acts as a stand on which users can store their current shaver, acting as a constant reminder to do their part for the environment.

Philip Thomas
Product Design

Spikey
Future concept for BIC

The BIC Spikey is a unique bottle opener that helps reduce the chances of drinks being spiked in public places. The device would be used by bartenders working in busy bars and nightclubs that are current hotspots for drink tampering. The product clips on top of glass beverage bottles and as the handle is pulled down, a sharp spike is forced through the bottle cap, piercing a hole. Once the handle has been retracted, the device is removed, leaving the cap still attached to the bottle. A straw is then placed through the hole in the cap and the beverage can be enjoyed with a reduced chance of being spiked. The BIC Spikey helps to uphold the company's new direction by being more responsible, both socially and environmentally. The device utilises the existing bottle cap rather than an additional plastic component to combat drink spiking.

Tom Jay
Product Design

As a company BIC have spent years producing products people trust, a product characteristic difficult to attain. While BIC have built these solid relationships with their customers they have neglected the social and environmental responsibilities associated with producing plastic products. Building on the brand trust already formed, the Portable Door Lock addresses the personal security involved in locking doors when on the move. BIC products can be used by anyone, anywhere and in a similar fashion this product locks any door, anywhere. With a simple ratchet strap pulling the two blocks together, it secures any door quickly and effectively. Made from bio-degradable plastics, when not in use the blocks ratchet together into single product to allow ease of transport when travelling around.

Michael Stanley
Industrial Design and Technology

Twist

A variable temperature kettle

Consumers are increasingly aware of the environmental impact of their decisions, but many are reluctant to change their behaviour. Technology brings improvements to people's lives, but some sustainable design neglects the user's needs, leading many to associate ecological living with compromise. Heating water requires a lot of energy; conventional kettles boil water to around 100°C, but often this is unnecessary. A cup cooling to drinking temperature is a common sight and brewing tea or coffee above 92°C can leave a burnt flavour. The Twist kettle enables the user to easily vary the water temperature, saving energy. The novel but intuitive interface allows users to choose the temperature by twisting the kettle. Its 360° base becomes a temperature dial, setting the temperature is an extension of an existing movement.

Alexander Ambridge
Product Design

The energy management system has two main components: an energy management console and a room power switch. Understanding user behaviours and attitudes towards reducing energy use was critical in developing the system. The console itself is portable and designed following inclusive and interactive design principles. Based on energy consumption targets set by the user, electrochromic ink is used for strong but unobtrusive visual feedback. Household as well as individual device consumption is displayed on the console; information users can then use to take action and reach targets set. Each room has a power switch next to the current light switch. When leaving the room or going to bed the user can conveniently turn off all devices in the room, reducing standby consumption and providing peace of mind for a good night's sleep.

Sam Verma
Product Design

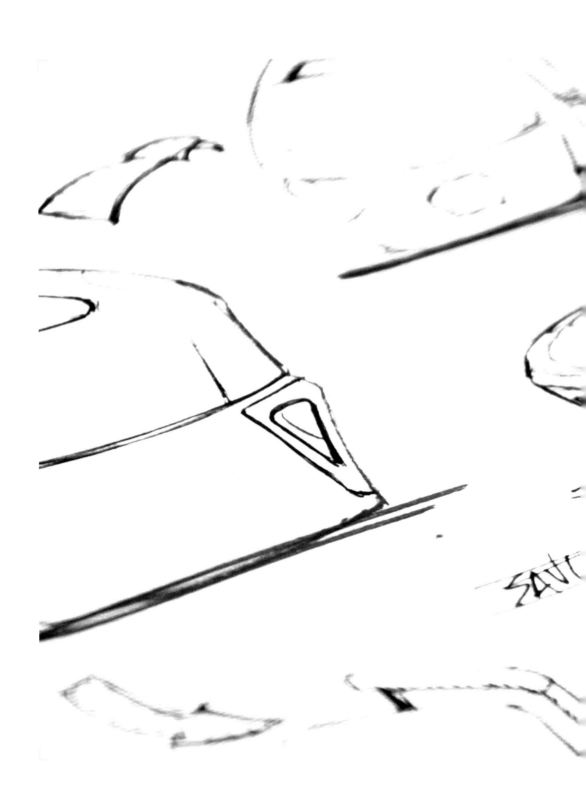

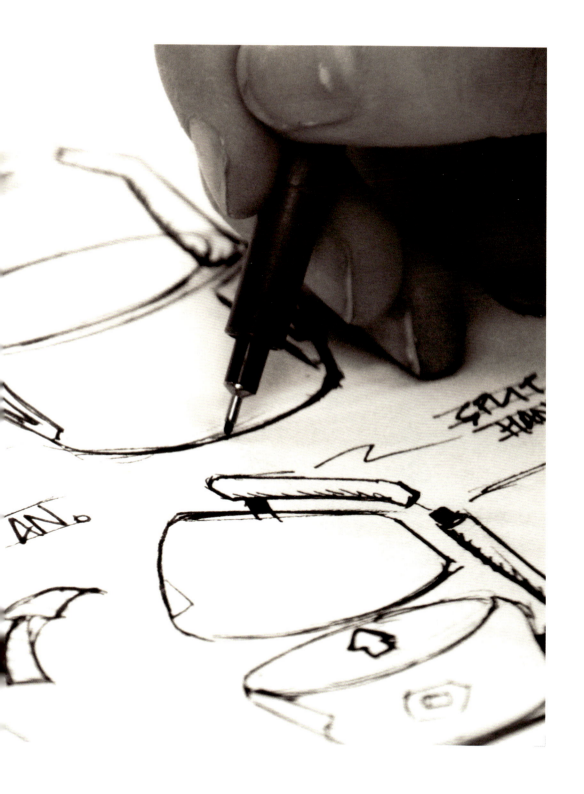

Waste Doctor
New supermarket service design for reducing food waste

Scan the barcode

↓

Food Alarm

↓

Food Doctor

↓

Food Village

Use-by date of Mature Cheddar is TOMORROW. Would you like to check the list?

No   OK

Freshness can Cost Less
A modulised energy saving refrigerator

The aim of this project is to reduce food waste by reminding customers of proper food management. The research highlights the fact that 8.3 million tonnes of food waste is produced annually by UK households, with expired 'use by' dates accounting for 60% of this. The premise of the project is that clarifying information on use-by dates can reduce food waste. After evaluating the range of behaviours, a new mobile application and supermarket system concept was suggested. The mobile application scans barcodes on a receipt, shares purchase data with supermarkets and offers a number of functions for managing an effective food plan. Users can receive reminders of approaching use-by dates (Food alarm), generate analysis of their personal waste patterns (Food doctor), as well as recommend and share food plans socially with other people (Food village). In the actual supermarket, customers can also use a digital scale (Smart scale) to calculate the correct portions of food needed and view an interactive store map showing sale offers regarding items in their food plan. The app has a game mechanic, with each service delivering points based on the individual's purchases, waste patterns and lifestyle. These will not only encourage the customer's brand loyalty but also expand business opportunities for the supermarket. Also it will motivate customers to plan food management more efficiently and carefully.

Refrigerators are one of the only appliances that work 24 hours a day, 7 days a week and as such account for a vast amount of energy consumption. Actually keeping food fresh can cost less. We just need a new refrigerator design. The modular energy saving refrigerator is designed to improve the energy consumption of refrigerators. It features customisable cabinets and the ability to control the temperature of each cabinet separately. With such features, people can have more flexible use of their fridge spaces. Users can expand the fridge space if they accidentally buy too much, or reduce the fridge space when there are only few items that need to be cooled. Users can also set some cabinets to 12-18°C to store some goods that need not be refrigerated, but are better kept at lower than room temperature, such as cosmetics. The glass door allows users to check their foodstuff without opening the door. Losing cold air through each door opening is one of the key issues that affect the energy performance of refrigerators. The proposed refrigerator fits in all typical UK kitchens and matches the size of other appliances in the kitchen such as ovens and washing machines.

Haeyoon Park

Ming-Chih Tsai
Integrated Product Design

Domestic Energy Reduction
A design organisation focusing on domestic energy use

A Heliotropic Energy Harvester
A passive solar tracing (heliotropic) energy harvesting system

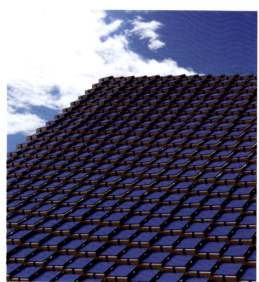

A design organisation that has two distinct yet reinforcing elements: the first follows the established activities of a design consultancy, creating physical products, systems etc.; the second innovative element is a video blog to inform behaviour choices, structured predominantly as a real world cooking show. The focus is on life in the kitchen because it is the energy nexus of the home, for appliances and the food itself (an often overlooked, but substantial, energy waste in the home). In this organisation, elements of the design process serve multiple uses, increasing output with minimal additional input. Take market research - identify a product niche, but also provide a consumer advice service. Use video ethnography to inspire product insights, but also to suggest more efficient behaviours. The product insight example is a kettle with an active water fill indicator to reduce over boiling, with an internal tricolour OLED, changing by designated increments such as 100ml or number of mugs. The behavioural insights, for instance clever washing up methods, are integrated into the cooking vlog. The vlog will be aimed at younger people living independently for the first time whose habits are still forming. To widen the target market, time and money as well as energy will be promoted as the major savings.

Solar energy is by far the most abundant ambient energy source on the Earth. However, solar energy only contributes to a fraction of the human energy consumption. The main reasons for this are the technological limitations and the cost of converting sunlight to usable energy. Increasing the use of solar power could help reducing CO_2 emissions associated with power generation. In this project, a passive solar tracing (heliotropic) energy harvesting system was developed. The system comprises interconnected units that tilt to face the sun, thereby increasing the solar collection efficiency of solar cells located on the units while eliminating shadowing effects. The tilting is achieved by a passive adjustment system surrounding the units, which based on bimetallic technology responds to changes in the Sun direction. Electrically conducting regions at the unit intersections provide means to transport the generated electricity for use and storage. Each unit can be made cheaply and the dimensions may be customised. However, the system is particularly well suited for smaller units, allowing an almost-flat solar tracing surface and a non-intrusive design. The system is suitable for roofs and facades, and is ideal for use in remote areas, for example in areas of the world where technical maintenance is not available. It is estimated that the system has up to 50% higher solar collection efficiency than conventional static solar panels.

Tom Wade
Integrated Product Design

Tor Sandén

The Lexi Bird Box
The Lexi Cinema one mile impact

The Lexi is an independent cinema located in North West London, in Kensal Rise. When they approached Brunel, the team was reviewing their advertising strategy and wanted to try something more sustainable and original than leaflets and posters. The project started as "something to mount on trees and buildings" and became the Lexi Bird Box. The box is branded with the Lexi colours and logo and is equipped with a solar LED that illuminates the logo at night to ensure maximum visibility. It can be disassembled in minutes without the need for screws, easing the cleaning and storage process. QR codes are printed on the box: when scanned with a smartphone, they direct the user to the Lexi website. Finally, the box is not just an advertising tool, but also comfortable nesting space for local birds!

Elena Bruno
Product Design

The Blue Badge Scheme is a system supported by the UK Government that aids certain categories of disabled population to travel and park. The number of issues facing the scheme leaves many users and local authorities frustrated for the lack of consistency, increased abuse by offenders, financial losses, violence and car vandalism often triggered by the current Blue Badge processing and design. The cause is not only the ease of obtaining and reproducing them, but the value these badges hold. The solution proposed is a shift from the current paper badges to an electronic system for quick authentification. This device is an attempt to counteract the high number of abuses, discouraging wrongful parking and badge theft, increasing productivity for parking attendants and guarding government costs on the long run.

Sorana Barbalata
Industrial Design and Technology

Sustainable Supermarkets: Tesco

Investigation into the design of sustainable supermarkets

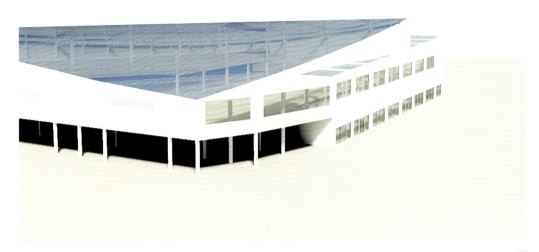

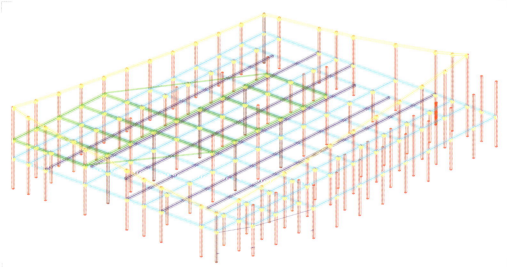

Using documents provided by a contractor working on a store for a national supermarket, the group focussed on the following four criteria: Drainage, structure, materials and management. The overall roof shape is designed to drain water effectively, while allowing natural light into the structure to illuminate both shop floor and staff facilities to provide a more pleasant environment for customers and staff whilst saving energy. The materials chosen emphasise the importance of low embodied carbon, having effective insulation to reduce the heat required for the operation of the structure. Sustainable drainage is incorporated in the form of a soakaway system. The whole structure retains the small footprint of the case study, so that other areas of the site may be used for residential buildings.

S. Brockie, R. McAvoy, I. Sidhu, A. Singh
Civil Engineering with Sustainability

Snapaway

A simple and effective way of holding four cans

Imagine a self imposed challenge. To devise a totally innovative packaging solution for transporting four beverage cans. Add to this brief: environmental impact, sustainability plus minimal use of material and manufacturing waste. Snapaway achieves this with an aluminium package that will be laser welded onto a beverage can ready to be torn away.

Joel Kemp
Industrial Design and Technology

Joe Midgley

Does Corporate Integrity Exist?

Companies are constantly trying to persuade us that they are the "nice guys". Whether its Starbucks boasting about their Fairtrade ethos or Levi's pushing their new WaterLess campaign, social responsibility surrounds us. It's undeniable that these ventures can be globally positive, but the motives are questionable. Big businesses aim to meet the expectations of their stakeholders and to deliver increased profit. Boosting one's positive image can work wonders, even haze over a wrongdoing. Coca-Cola's Indian water debacle is a prime example. The company was accused of extracting groundwater causing severe water shortages for the local community and putting thousands of farmers out of work. They saved themselves by highlighting a water stewardship policy in a sustainability review using a good deed to bury a bad one… and their explanations were plausible. The digital age is increasing corporate transparency; it's no longer easy to cover up mistakes as people are becoming increasingly perceptive about big business. The 'Anti Corporate Movement' is growing in strength; consumers are becoming increasingly savvy and are less likely to believe superficial claims. So what does the future hold for social responsibility? And are there any companies out there who care for the right reasons?

Corporate Social Responsibility (CSR) was born through the realisation that the area affected by a business stretches much further than the customer, stakeholders exist in all kinds of unexpected places. It began as an inclusion of public interest into corporate decision-making and the honouring of a triple bottom line: people, planet, and profit. In 1980, almost 100% of a company's assets were defined as tangible. Today this figure is between 30 and 40%, the rest is intangible, much of it being the company brand. The brand is often the biggest asset a company owns. The resulting worth of the brand encourages modern companies to invest a great deal in their image and reputation, with any mishaps being potentially disastrous. In April 2009 when the Innocent Drinks brand received a £30 million investment from Coca Cola, their YouGov Brand Index "Buzz" rating slipped from five out of seven to one. The niche smoothie brand was seen as "selling out" and its popularity dropped, highlighting the influence of how a consumer perceives a brand. Today over 90% of the Fortune US 500 companies employ CSR initiatives and the use of CSR is an effective method of shielding firms from scrutiny and allowing them to trade more freely with consumers and stakeholders. It is important to explore the CSR techniques employed by firms and the ways in which they can be somewhat misleading.

A common catalyst for increased social responsibility tactics is the global concern over the environment. In Apple's environmental report, they state that they recycle 66.4% of the materials used in their products, but how do we know how much of this is materials from defective products simply being reprocessed? The way in which many large companies have developed an eco-presence in a relatively short period of time is a further reason to be sceptical. The 2010 BP oil spill resulted in the death of eleven individuals and 200 million gallons of oil being spewed into the Gulf of Mexico and this dramatically shifted people's belief in them as an environmentally caring company. Plugging the leak was the primary concern, however BP spent valuable time and money spraying toxic dispersants on the ocean. These dispersants did nothing to reduce the toxicity of the oil; they simply broke it up into tiny droplets, therefore reducing the visual impact of any photos taken. In July 2010 they were forced to admit using Photoshop to exaggerate the level of activity in the Gulf oil spill command centre, the photo showed staff monitoring ten giant screens, in reality three of them were blank. These incidents portray the company as desperately trying to enhance an image of themselves, which did not exist. Three months after the Deepwater Horizon rig exploded, it was revealed that BP was yet to update their oil spill emergency plan. Errors included the listing of a wildlife expert who died four years before the plan was approved. All of these factors portray BP as being more concerned about retaining an environmental image, than actually being "green".

To be successful, CSR must be fused into the fundamental ideals of a brand; it is not a tactic that companies can employ as a 'quick fix' for a delicate situation. Modern consumers are becoming increasingly savvy and can spot the cracks, which inevitably show weak social responsibility. The BP situation was clearly worsened by the fact that they had so overtly attempted to portray themselves as a 'nice' company, developing an 'eco' logo and marketing a positive slogan isn't enough. It is true that bad news travels quicker than good news, often being broadcast through numerous online media such as Youtube, Facebook and Twitter. This is exactly what happened after the BP disaster: an internet blogger used Twitter to mock the energy giant by pretending to be BP's Global Press Office. He gained more than 135,000 followers, showing the power of the internet by enabling one man to make a laughing stock of one of the world's largest energy companies.

Although transparency makes businesses uneasy, it also offers opportunities. If a profit-making business begins with honourable aims, then surely it is possible to for them to continue in that way. Blake Mycoskie began his profit-making business after witnessing the struggles faced by charities giving shoes to people in Argentina. Mycoskie explained his reasons for creating TOMS shoes, stating "I really wanted to attack the idea of predictability and sustainability". By creating a profit business, which donated one pair of shoes for each pair of shoes sold, he would be able to prolong the donations. This business was founded with wholly good intentions and by maintaining its simplicity and transparency has been able to grow in an honest and successful way. "There are no formulas or percentages, you buy a pair of shoes, and we give a pair of shoes". By retaining their goal, this business has actually made a real difference to people's lives whilst enjoying rapid growth and healthy profits. There are no hidden details or embellished facts, they have never even launched a traditional advertising campaign, Mycoskie explains that the customers do this for them by speaking positively of the company. They also organise monthly trips, where volunteers can become part of the "shoe drops", further highlighting the honesty of the business model as well as the brand.

It is likely we will see a shift in the value of corporate business back towards the tangible assets it produces. Consumers are becoming very brand intelligent; these 'Prosumers' know what they want and aren't likely to be tricked by hollow marketing campaigns. Generosity and compassion are becoming increasingly important. Muhammad Yunus captures this ideal; "Once poverty is gone, we'll need to build museums to display its horrors to future generations. They will wonder why poverty continued so long in human society - how a few people could live in luxury while billions dwelt in misery, deprivation and despair". People have never been so aware of the selfishness, which has existed for so long in the corporate world, and its time for an honest change.

Future Concepts for Tata?

For 140 years Tata have maintained a strong sense of heritage in the values-based approach to business they practice, and as a major operator on the global stage subject to increasing public scrutiny, it is vital they maintain their position of responsibility. In 2020, issues of social and environmental sustainability will be ever more familiar and accepted. These products depict Tata's role in delivering positive solutions to these challenges whilst meeting people's essential needs. As consumer touch points within larger function-led systems, they provide an embodiment of the Tata brand in both local and global contexts. With an emphasis on creating a common identity across disparate sectors, this new generation of products addresses the aim of brand unity, with designs focusing on longevity, durability and simple user interaction.

Alternative Battery Cell
Future concept for Tata

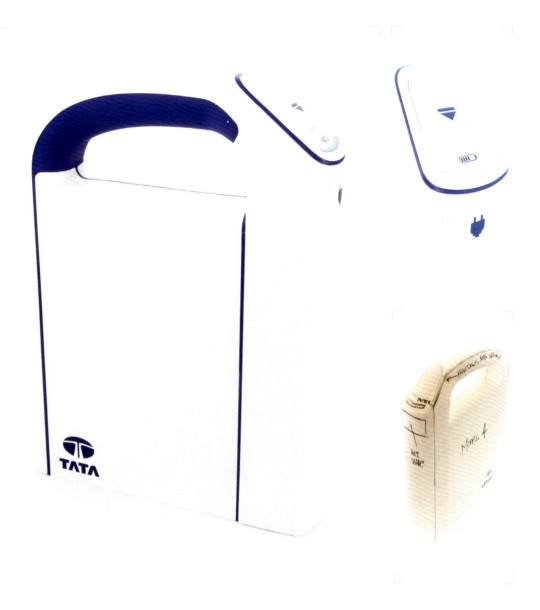

Distribution and storage of power affects millions of households without reliable access to electricity. This concept battery cell uses new technology developed by MIT to replace the anode in traditional Li-ion batteries with carbon nanotubes, resulting in ten times the energy density of existing products, incorporated into a package that is highly portable. The cells themselves are designed to be easily removed from the casing and replaced when they lose capacity. The product is not only restricted to developing nations with intermittent power, with the growth of electrical vehicles also hampered by the high weight-to-energy ratio of existing batteries. A single energy cell would power a small car for the journey home from work and could be easily swapped out.

Buster Palmano
Product Design Engineering

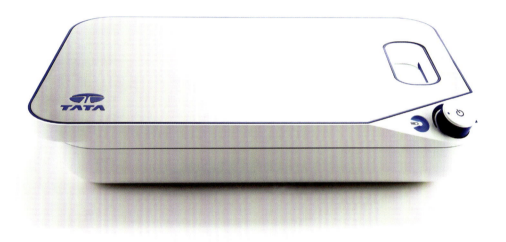

Through careful investigation into the world of 2020, it was possible to highlight areas across the globe where TATA is capable of dramatically improving people's lives. In particular the developing nations across the world which are commonly subjected to an exploitation of household fuel. To empower the people subject to this exploitation, the concept to provide a source of heating in the home was developed.

This heating source is provided through the use of induction, transferring the dependency from the exploited natural fuel to a sustainable electrical energy. The Cold Stove is capable of providing a mobile cooking plate, a room heater and will also provide a means to heat water. This minimises the dependency in the home to one sustainable product capable of supporting three different heating purposes.

Harrison Williams

Product Design

Voltas CC

Future concept for Tata

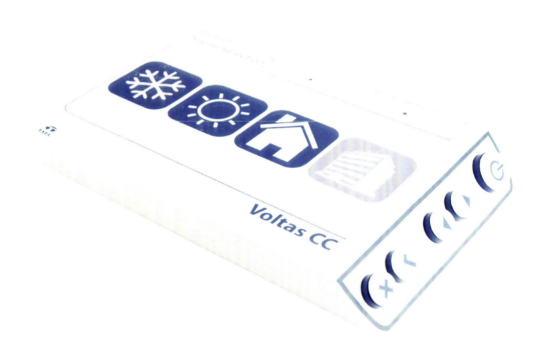

It has been suggested that future AC units from Voltas will implement new "Thermoacoustics" technology, which uses sound waves instead of traditional chemical refrigerants to cool and heat the air, dramatically reducing energy consumption. Additionally, AC units will focus on user interaction, aiming to offer the user a greater sense of control and empowerment. Voltas CC is a control panel that wirelessly manipulates any of the new Voltas conditioner units, using Bluetooth technology. The panel is a component that, unlike existing Tata units, provides users with simple controls and interface options, enabling users to more simply manipulate any unit. Instead of detachable remotes or hard to access controls on the actual AC unit, the panel will be wall mounted allowing the user to locate the device where they feel is convenient.

Peter Williams
Product Design

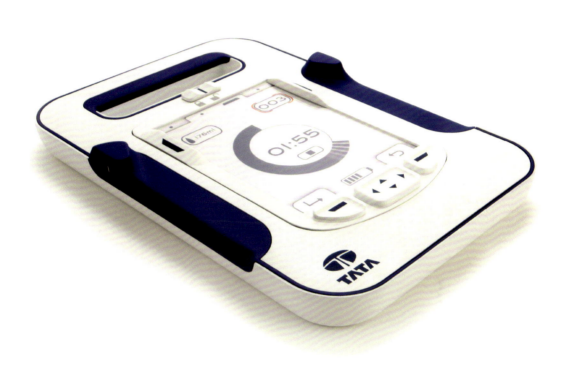

Attempts at managing solid waste are outpaced by generation rates, with landfill now widely recognised as unsustainable. In developing nations, resource scarcity often stifles progress in the improvement of quality of life. Continuing a legacy of active social responsibility into 2020, the waste processor concept addresses these issues, enabling more effective resource utilisation whilst offering a solution to a growing waste crisis. Within it, a modular system of catalytic pyrolysis reactors process hydrocarbon oil from plastic-based waste streams, including packaging and e-waste, outputting fuel and polymers. As Tata's primary touch point for local operators, the control interface enables simple scaling of operations, empowering people with the tools to create real value and build resilience within communities worldwide.

Jermaine Legg
Industrial Design and Technology

Tata Water Bottle

Future concept for Tata

Climate change is introducing new uncertainties into water resource management due to erratic and unpredictable rainfall. India currently possesses 16% of the world's population but just 4% of its water resources, by 2020 the major cities will be unable to supply the population's needs. These cities are already struggling, as currently only 74% of the urban population have access to a basic water supply each day. Tata is in a unique position to influence changes needed to create an improved water supply, as the company is already heavily invested in this area. The system collects water during monsoon season and stores it ready for drinking. Using existing Tata technology the water will not need to be boiled or UV filtered. By creating an accessible public water harvesting point, people will be able to collect water using their Tata water bottle whenever they need.

Madeleine Carver

Product Design

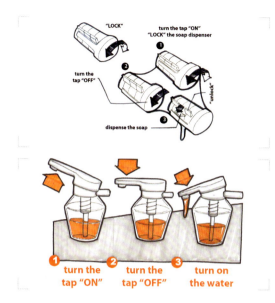

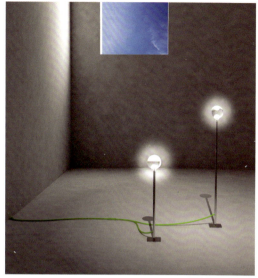

Domestic water usage has increased dramatically. Showering is a large contributor, which also uses high levels of energy consumption in heating water. From the persona that was developed during primary research, it was found that whilst showering, people tend to keep the water running whilst soaping. However, this action is not actually necessary because, when standing under the water, the soap will be washed away quickly. This behaviour is therefore motivated through a different level of human needs as opposed to purely cleanliness, as there is no hygienic advantage in leaving the tap or shower running. The project looked at how this aspect could be addressed in order to encourage users to use less water whilst soaping. The proposed design integrates the shower tap with the soap dispenser to make people intuitively stop the water before getting the soap. It offers a new way of controlling the water and soap together without drastically changing the wider behaviour of the users. When the tap is in the central position, the system is off and no water will flow. The tap is then raised to turn on the water to wet the user. In order to dispense the soap, however, the tap must be pulled down, stopping the water first, and then depressed further to dispense the soap. Finally, the water is turned on again to wash the soap off.

Green-light is a novel lighting solution that will change people's perception of lighting. It is a hybrid lighting system composed of an LED source, and a piping system to carry natural light from solar collectors. Combined with sensors, Green-light would be an effective sustainable lighting solution. During the day, a sunlight collector placed on the roof or balcony directs the light to any desired interior space using optical fibre piping collects. The organic shaped light bulb will glow, giving out a soft, warm light source for daily activities and indoor plants. When there is not enough sunlight, LEDs will light up to compensate for the loss of brightness. The main advantages of Green-light are sustainability, energy efficiency, lower electricity bills, flexibility, and hygiene. Green-light can be used in both domestic and commercial environments. It brings sunlight to rooms with few windows, buildings in the shadow of their neighbours, and rooms with north-facing windows (less sunshine). The installation cost is relatively small for this system when implemented in the build phase of new properties. Energy saving with this lighting system is very effective when the architectural space is large, such as in commercial buildings.

Carson Kan
Integrated Product Design

Sze Yin Kwok

WaterBox

The optimisation of a reverse osmosis water filtration device

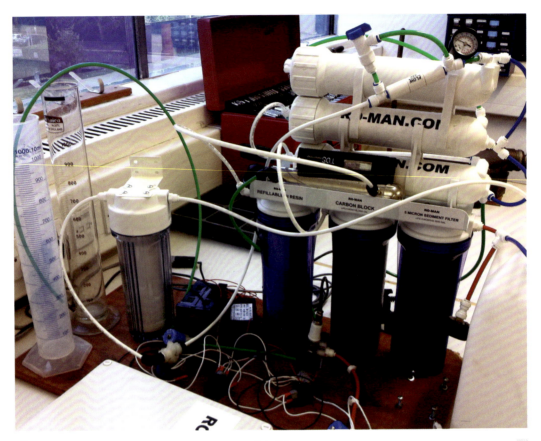

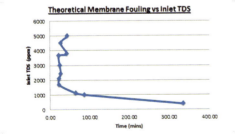

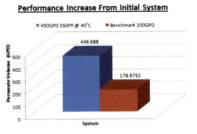

The project aim was to significantly improve performance of an existing product previously developed at Brunel University known as the WaterBox. WaterBox was designed for producing potable water for disaster relief situations. This project investigated the characteristics of the various system elements in order to optimise its performance and further develop the technology. Theoretical calculations and a series of experiments to assess the system's performance are to be carried out for a range of operating scenarios. By identifying and predicting performance trends, an optimal design can be implemented and tested that is capable of producing higher quality clean water, and/or a higher volume of clean water. Further research will identify the specific target market for this product.

Philip Jones
Mechanical Engineering

A modular device for a conservation-tillage cultivation system

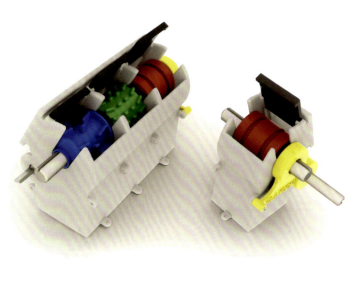

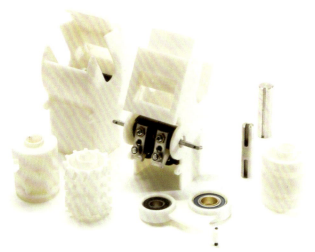

Sustainable agricultural systems reduce labour, fuel and machinery costs while improving soil structure and profitability. There are also environmental benefits, such as carbon sequestration potential which reduces carbon dioxide and nitrous oxide emissions. The focus was the design of a seed metering system tailored for a direct-sowing application, maximising benefits for farmers and the environment.

In collaboration with Claydon Yieldometer, a UK-based seed drill manufacturer, this system offers a reduced part-count, and three interchangeable metering wheels specifically designed for metering oil-seed rape, cereals, pulses and granular NPK fertiliser. The modular design allows the system to be optimised to exact capacity requirements, offering versatility and adaptation currently unmatched.

Dan Smith
Industrial Design and Technology

Radial Wind Turbine

Turbine to accept wind from any direction and be easy to assemble

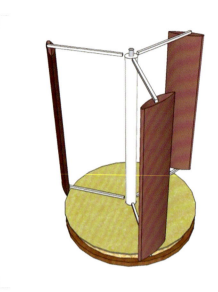

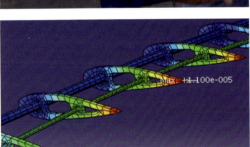

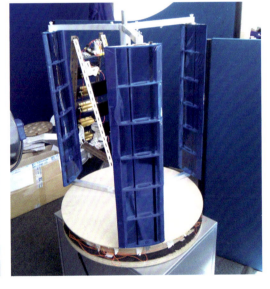

The aim of the project was to create a portable radial wind turbine that was optimised using CFD, had a suitable generator and was easily assembled from the storage configuration. The final design was a vertical axis wind turbine with three blades connected to a hollow central rotating shaft, with a base housing the generator and stability from a fixed central shaft. The main advantage of this design is that it can accept wind from any direction, unlike the more common horizontal axis wind turbines. The design was optimised to obtain the best aerofoil for the blades and to find the connection angle, as well as ensuring that materials and dimensions were appropriate. The turbine breaks down into six parts for transportation or storage and is easily assembled using clips and brackets to lock various parts into the operating configuration.

J. Bennet, K. Hassell, M. Rahman, S. Sadiq, S. Sekar, S. Subramony

Aerospace Engineering

The Brunel Solar Powered Aircraft is an Unmanned Aerial Vehicle designed to increase flight time using 52 solar panels within the wing. The experimental aircraft also contains a payload of surveillance equipment, lending itself to applications such as military, wildlife conservation, environmental monitoring and cartography. The aircraft is based on a typical glider design, with a 3.5m wingspan and a V-Tail stabiliser configuration. The design combines wrapped and pultruded carbon fibre; along with balsa, ply and spruce to form the main structure. The aircraft has a cruise speed of 10m/s and a maximum speed of 18m/s. The prototype aircraft has been successfully flown supplemented by solar power, while ground tests indicate flight time can be expected to increase threefold with solar power.

S. Blakeney, J. Palmer, T. Shadbolt, J. Simpson
Aerospace Engineering

Volans Eco
Autonomous luxury yacht

EL-1
Low cost electric tricycle

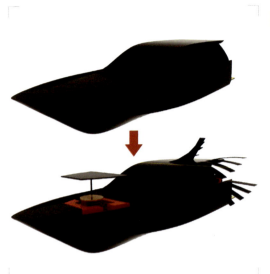

Through the use of autonomous technology as a design opportunity in the field of yacht design, it is possible to focus on sustainability, function and aesthetic appeal. The design incorporates various solar panels including a large area of the body constructed entirely of photovoltaic panels. The fins can be closed as the boat is navigating in order to reduce air resistance and when anchored the same fins can be expanded and the photovoltaic surfaces rise to face the sun and follow it to recharge the onboard batteries. This innovation means that even in areas of partial shade the Volans Eco can continue to harvest solar energy in an efficient way; increasing yield by an estimated 45% more energy than fixed solar panels. The solar-powered yacht is made from fibreglass and includes enough photovoltaic panels to generate about 8KW of renewable energy per hour. The aesthetics of the yacht exterior aims to make the customer perceive it similarly to a top-class yacht rather than an eco-friendly variation whilst naturally reducing the negative effect on the environment. The yacht has been based on the desire to combine an aesthetically competitive form with the use of photovoltaic modules.

Technological development has significantly increased the capability of electric engines; this will be a big trend in the developing countries as electrical energy is cheaper than petrol or diesel. The EL-1 incorporates the use of electric engine, which will provide a cheap and ecological transport. EL-1 is an affordable electric tricycle that features higher safety performance than a conventional two-wheeled vehicle. The EL-1 is aimed at developing countries where people use motorcycles as their primary means of transportation. Motorcycles are very popular in these countries because they provide a cheap effective personal transportation solution. In some countries people use their motorcycles to carry goods with, or even use them as taxis. The roof assembly is a safety feature, and acts as a roll cage to protect the user in a collision, which will provide far greater safety than a motorcycle. For convenience there are two storage areas, one is under the seat and the other is the luggage at the back. This project focuses on lowering the cost of personal transportation due to the low income of populations in developing countries. The design essentially considers minimizing the number of components and reducing material usage. The form is simplified to reduce the cost of moulding, which in turn lowers manufacturing cost.

Ivan Yang

Noppan Kaewkanjai
Integrated Product Design

Sustainable Campus Bus Stop

A self sustainable solution through the use of solar energy

Making Something Last

Sustainability through longevity

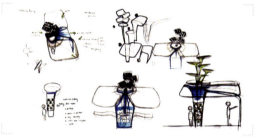

The final concept is an intelligent self-sustainable campus bus stop. Aimed at reducing environmental impact, it converts solar energy into electricity, collects rainwater in order to grow grass around the bus stop and evokes the feel of nature, supplying the most convenient and comprehensive new experience to the users. The concept improves campus living by redesigning Brunel University campus bus stops, supplying a new service to students. The bus stop could be a landmark building emphasising Brunel's innovative reputation and enhancing users' psychological experience. It is important that the bus stop is well integrated within its surroundings, making the architecture come to life in the academic atmosphere. The campus bus stop is a good place to evoke lifestyle thinking and improve students' environmental awareness, encouraging behaviour change. The concept will require innovative design techniques to reduce the environmental impacts though sustainable design. It will create an environmentally friendly artwork, not only a functional bus stop.

By extending the life of products through design, we will create better products, allowing the consumer to be more considered in purchasing, using less resources and thus enjoy a better quality of life. Growing awareness that we are rapidly using up resources and cannot consume material goods in the way we have in recent years, combined with the impact we are having on our environment, is encouraging new types of consumers and businesses. This range has been designed using natural materials, to appeal to the "green consumer" and their ideologies, together with businesses that have sustainable strategic aims. This modular furniture system encourages the consumer to choose the components that meet their needs today and then add to them as their requirements change throughout their life, every component being interchangeable. The "Table for Life" range includes a choice of short, tall, and personalisable bases. The surface comes in a range of sizes and has recesses that can be used as a temporary repository or to display items that the owners identify with themselves. Personalisation is seen as one route to postponing product replacement by increasing product attachment. Throughout the design process key influences have been: understanding the consumer, reflecting the design philosophies of past designers of long lasting designs, and promoting product attachment.

Yupeng Meng
Integrated Product Design

Penny Bamford

Can Fashion be Ethical and Sustainable?

Investigating design, ethical and sustainable strategies for fashion brands

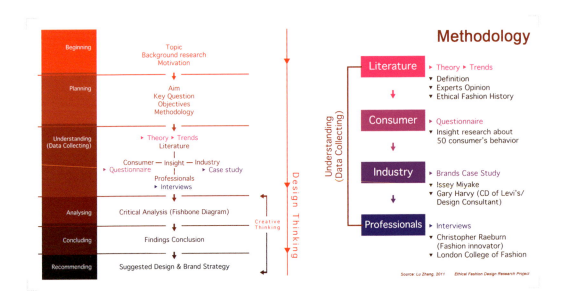

Methodology

Source: Lu Zhang, 2011   Ethical Fashion Design Research Project

Fashion today is fast moving and a complex mix of trends, phenomena and people. Both ethical and sustainable fashion exist but will these trends emerge as the basis for a future fashion design movement? Fast fashion is not just an impression, but it is a basic business model. It has changed shopping trends throughout the world, resulting in more people buying more clothes, more often, often cheaper, and ultimately wasting them or at least not realising their full wear potential. The fashion writer Sandy Black states, "In the last 15 years, fashion has become faster and cheaper." Should we start to rethink the wasting and polluting ways of the fashion industry? During this research, ethical/sustainable fashion brands were identified but ethical fashion still has a long way to go to reach the 'high street'. This research project is to create a new design-led brand strategy for Christopher Raeburn to improve its brand vision. It is a potentially ethical fashion brand, which creates innovative street wear using re-appropriated military fabrics. The research methods use both desk research and primary research from four aspects: literature,

> ## "In the last 15 years, fashion has become faster and cheaper."
> Sandy Black

consumer journey, industry and professionals. One professional, Dr Sharon Baurley has experience with research in ethical fashion design. In the Brunel public lecture – Living in a Wearable World, she states "Fashion trends do create waste, but smart textiles and materials could facilitate a more sustainable future where garments can be updated and changed. The development of ethical fashion is not being ignored but are the brands paying sufficient attention? Being sustainable could be a main trend and differentiator in the fashion world of tomorrow. Although there are a number of ethical/sustainable fashion brands today, they fail to connect with a wider base of customers and do not deliver a broad enough message on the value of ethical fashion in the future. Research to date suggests that increasing the customer experience will create a stronger understanding of the ethical and sustainable elements that can play a part in designing a fashion brand strategy. Sustainable designers believe that every little thing we bring into the world today will change the world of tomorrow.

Lu Zhang
Design and Branding Strategy

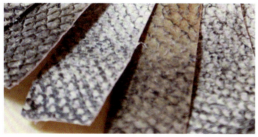

The Atlantic Salmon is an extraordinary fish that can swim thousands of miles to return to its original breeding water. Its anatomy is perfectly adapted for survival in extreme conditions and its skin has been found to have some exceptional properties. The fish featured in this project originate in the Northwest Highlands of Scotland. They were reared in an environment with a strong commitment to marine conservation, animal welfare and sustainability. A beautiful, exotic and unique material can be produced from salmon skin, but only when it is treated with the greatest care and respect at every stage of its development. A specially adapted sulphide and chrome-free tanning process has been developed to create this leather which has also been made into an exclusive sample product by Mulberry.

Unji Moir

Product Design

Future Concepts for Wrangler?

Flash floods, hurricanes, tsunamis, earthquakes and wildfires are common, but devastating, occurrences. The people who are worst affected are often urban dwelling civilians with no concept of basic survival skills. These disasters strip people's lives down to a primitive level and their daily struggles become based around primary needs, such as how they will be sheltered. Wrangler embodies the concept of the true survivor: the individual who is able to meet his or her needs through knowledge and raw physical capabilities. Wrangler empowers survivors of natural disasters by providing them with the necessary tools they need to begin rebuilding their lives.

The Wrangler Compass
Future concept for Wrangler

Recent years have seen a shocking increase in the occurrence and severity of natural disasters. Flash flooding in over populated developing economies has forced communities to live in areas at high risk of flooding. This project predicts that with an ever changing climate these natural disasters will only increase. One of the particular issues that victims of flash flooding experience is that they become isolated and their ability to find emergency help is impeded. The navigational device empowers their survival; allowing them to locate and navigate to emergency help points that distribute food and medical supplies provided by their government or through international relief aid from charities and other governments. It provides information on the types of aid being distributed, the amount of aid available and on crowding.

James Clarke
Industrial Design and Technology

The Wrangler Axe is adaptable to many post natural disaster scenarios and has been designed to meet high user expectations. Under the tanned aniline leather protective cover is the 2.5 Lbs hardened carbon steel axe head that allows effortless cutting of wood. The axe head can be used as a square allowing adaptability of the product to situations where accuracy is key. The Hickory handle allows the user to effectively cut wood safely and provides a comfortable rest point between strikes. The idea of the user not having to put the axe down to rest was inspired by what the axe represents; a chance to rebuild their lives. By allowing the user to be able to rest without putting the axe down, things are looking positive and life can be restored through their own intervention and initiative.

Ross Dexter
Industrial Design and Technology

The Wrangler Bow Saw
Future concept for Wrangler

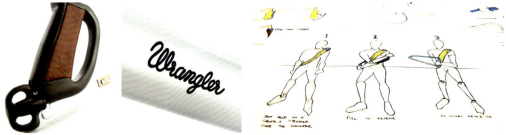

This saw empowers the survival of hurricane victims by aiding with the redevelopment of damaged areas through re-utilisation of debris materials. It's used for cutting timber as well as fallen tree trunks and branches. It's likely the user will be required to climb a tree or manoeuvre through an awkward situation to use the saw; it has therefore been designed with portability in mind. The steel frame contains a nylon strap which unwinds using the same mechanism as a traditional tape measure. The strap is unwound from the rear end and clips into the front of the saw frame; this allows the user to carry the saw on their back. It's partnered with a cover to prevent injury from the hardened steel blade during transportation. The polypropylene handle features a padded aniline leather touch point for increased durability and comfort during use.

Joe Midgley
Industrial Design and Technology

The Wrangler Mask responds to the post-natural disaster effects of bushfires. The respiratory conditions exacerbated by the smoke and haze produced from fire results in hazardous conditions. The mask aims to prevent health risks by providing residents with vital inhalation protection from toxic fumes and dust. High risk areas will be automatically supplied with the mask during a residential meeting that highlights the vulnerable areas of the town/city. Those caught in sudden disaster will receive the mask from fire fighters dispatched across the area. The mask sits within a protective leather pouch that slips neatly inside a wallet, purse or pocket. The mask folds out revealing two ear straps that extend to attach to the face and offers maximum performance of up to three uses.

Samuel McClellan
Industrial Design and Technology

The Wrangler Spade
Future concept for Wrangler

This concept for Wrangler is based on assisting people who have been affected by natural disasters, specifically those affected by earthquakes. By allowing them to clear away rubble, debris and other obstacles from communal areas, it will help them to resume normal life more quickly. The Wrangler Spade will provide the basis for the reconstruction of a community, empowering survival through a trustworthy, reliable and rugged product. This is a high performance earthy product that can always be depended upon. After the initial clear up, the spade can be used for other everyday tasks from gardening to building, never to be rendered useless.

Bradley Wherry
Industrial Design and Technology

The Future of Water Supply System Design in Rural Africa

A multifunctional design for community economic and physical well-being

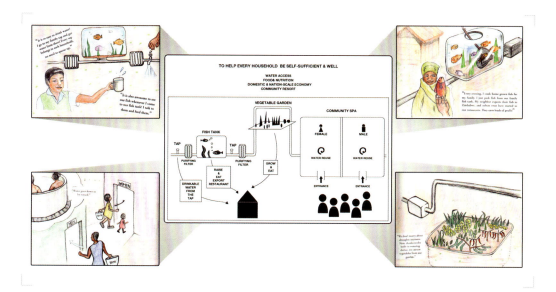

Design thinking was the starting point to future forecast the African rural regions' problems. The motivation for this project is the growing African population without access to clean water, resulting in malnutrition and other diseases. Scientists predict that, by 2030, desertification will further affect food security. At the same time, rising temperatures in Africa's large lakes may erode fisheries. The outcome of this project is a multifunctional system design that can be applied (in 2030) to help the rural African people's physical well-being and economic independence. This solution was approached through creative problem solving rather than focusing merely on visual creativity. Kevin McCullagh defined problem solving as "analytical interpretation of the current world." Designers should have knowledge of the problem domain, use analytical thinking and bring innovation in today's context. The methods we used include contextual research, hard, soft and mega trends, expert opinions, blue-sky and scenario generation. First, future research was begun by doing contextual research investigating the location, economy and socio-cultural factors. Investigation was made into hard trends in African population growth and the countries' economies including their volume of the crop production and economically active population, while also searching for soft trends

chronologically in order to study existing solutions and their technologies. From this trend observation it was found that evolving technology coupled with design thinking could lead to an innovative futuristic water supply system that would offer new opportunities to rural African communities.

Health and environmen was then extrapolated from the mega trends towards 2030 and converged into the future design. This type of trend exploration is also emphasised by many Brunel future design presentation speakers, such as Simon Black (DesignBridge) and Richard Walzer (NewEdge): "It provides rationale for innovative futuristic strategies". In line with our team's future design vision within the system, Professor Heinz Wolff supports science and technology to be the key drivers of future innovation, contributing to the "reestablishment of the social and financial sustainability". Wolff also sees future innovation as "renovation", meaning iterative innovation in technology. Emphasising his dual concept of 'frugality' and 'mutuality', Wolff believes that "technology needs to be attached to people" in order to meet the future challenge. This blueprint may seem like a grand utopian idea but it is firmly believed that current technology is still evolving and with design thinking, it can greatly contribute to the life of every human being.

Y. Chang, Y. Chien, S. Hsu, S. Jiang, J. Lee, Y. Liao

Design and Branding Strategy

Podoconiosis Prevention Shoe

A cheap and durable shoe to be made in Ethiopia

Podoconiosis affects millions of people, it is a swelling disease of the lower limbs that can limit a person's ability to work. Within many cultures this disease is misunderstood leading to people being outcast from their communities. It is caused by prolonged exposure to volcanic soil. It is found in developing countries including Ethiopia, where many people cannot afford a pair of shoes which would eradicate the occurrence of the disease. This shoe design is made from local leather, the design is uncomplicated and made from one single piece of material which ties to itself. Using a single material with no joins the final shoe's costs are kept extremely low. The shoe can be flat packed and distributed easily to the most needy. If the shoe is worn the chances of the user contracting the disease will be eradicated.

Tom Collett
Industrial Design and Technology

Street food vending conditions in Ghana are a huge cause for concern. Toilets and washing facilities are scarce and the washing of hands and crockery are done in buckets or bowls. In December 2010, extensive research was carried out in Ghana giving a huge insight into the industry. Poor hygiene practice was not just the result of bad cooking conditions, it was a matter of behaviour. The aim was to improve bad hygiene practice with the possibility of economic growth. The Mukase is an apparatus for the chopping of meat and vegetables. The kit also has a soap dispenser and a tray with a drainage system that filters used water. Mukase was designed to encourage the washing of hands whilst cooking, but also to be remade with accessible materials and simple tooling, empowering local carpenters to make it themselves.

Efua Mensah-Ansong

Industrial Design and Technology

Innovative use of manufacturing, materials and electronics.

humanistic innovation   19

sustainable innovation   197

technical innovation   245

digital innovation   321

network directory   353

Food Hygiene and Safety in a Commercial Kitchen Environment

Ultrasonic atomiser to eliminate harmful transient and resident pathogens

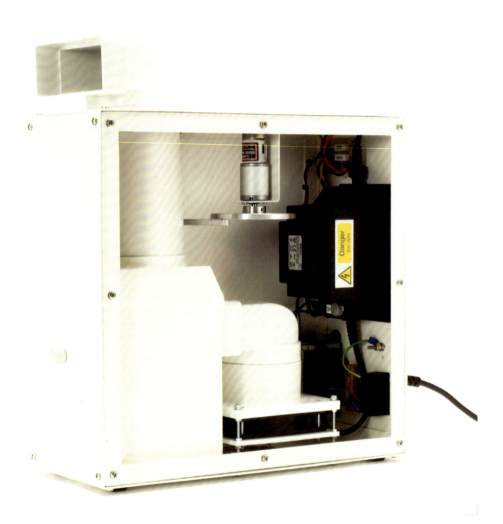

In spite of extensive teaching on food hygiene for the prevention of food-borne disease, the incidence of outbreaks and sporadic cases continues to increase. Evidence suggests this is a result of the neglect of basic food safety principles. 97% of people in the UK want eatery hygiene ratings to become available, highlighting an increasingly health conscious Western society. Colloidal silver, an ultrasonic technology, is used. A natural antimicrobial agent is diffused via distilled water into the kitchen's atmosphere, eradicating harmful resident and transient pathogens from the air and surfaces. Ultimately, this device does not eliminate the need for the physical removal of 'soil'; it complements the processes involved in disinfecting the kitchen whilst unoccupied.

Stuart Wickens

Industrial Design and Technology

Stirling engines are extremely efficient heat engines which run purely off a temperature difference. Despite these fantastic qualities, their adoption in industry has been very slow. This project focuses on applying them to a third world environment to pump safe drinking water to deprived villages using a parabolic mirror to capture sunlight. A variable phase angle mechanism allows the engine to optimise itself for different temperatures automatically throughout the day without the use of electronics. This significantly increases its pumping capability, making it a much more viable option for the future.

James Willson
Industrial Design and Technology

Wing Fan
An efficient and affordable ceiling fan

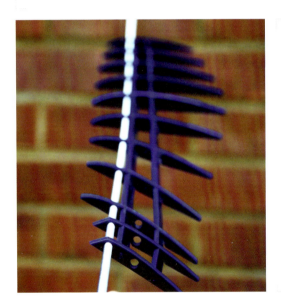

" 129% more efficient than competitor products, costing 50% of an air conditioning unit over its lifetime"

With a design unchanged for 130 years, ceiling fans are used in the US and Far East by those unable to afford air conditioning. Developed in collaboration with Tom Dixon and DesignPlus, the Wing Fan addresses many of these problems, without the dryness created by air conditioning. The flat blade and fixed pitch of existing fans was replaced with a twisted aerofoil shape to vastly increase airflow. Flared wingtips observed on soaring birds and a secondary wing present on waterfowl were adapted to further increase the efficiency and reduce the noise of the blades. Gains were computed using fluid dynamics and verified with physical prototyping, with the resulting fan 129% more efficient than competitor products and costing 50% less than an air conditioning unit over its lifetime, while maintaining a striking aesthetic in line with the brand.

Buster Palmano
Product Design Engineering

The commercial design brief was set by a packaging based consultancy, iDi Pac Ltd a packaging based consultancy, to create a dispensing unit for soluble based granular chemicals. The packaging product had to have the capability to adjust the dispensing volume to the user's requirements whilst being adaptable across several markets and industries for wider spread of the concept. A focus on the instant coffee industry was drawn upon to give direction to this broad brief. Using instant coffee, human-factor based processes were used to define the needs and errors of the problem for users. Once this was established the concept was built upon to be adaptable for other granular bulk materials by gaining a deeper understanding of granular flow systems.

Tom Jones

Industrial Design and Technology

The ATD Gun

A human-centred Adhesive Tape Dispenser

Through communicating closely with an online bookstore in Cambridge, it was determined that the packaging department struggled to work efficiently with the current market tape guns. In some cases it is even causing physical harm through repetitive strain injuries. The danger of economic impact through law suits could cause many small to medium sized enterprises to file for bankruptcy. The ADT Gun looks to alleviate the physical degradation of the user. Functions have been designed to remove the hyperextension and compression of the wrist. Secondary to this, a new means of cutting the tape post-application has been designed to remove the force needed to snap the tape.

Harrison Williams
Product Design

The ultimate luggage, developed in collaboration with international product development company PBFA, makes regular business travel easier. The Suit Wrap allows a suit to be packed, arriving crease free. Utilising a novel two-part system, the suit is kept neat, but becomes easily portable. The Suit Wrap also protects all other essential belongings for a business trip, including easy access pockets for a laptop and toiletries. Key to ensuring that the case is able to meet the requirements of the customer are the considerations that make the case simple to pack, wrap and wheel. The Suit Wrap is aimed at saving the regular traveller time and hassle on those stressful business trips.

Chris Strickland

Product Design Engineering

Boot Trolley
Collapsible shopping trolley

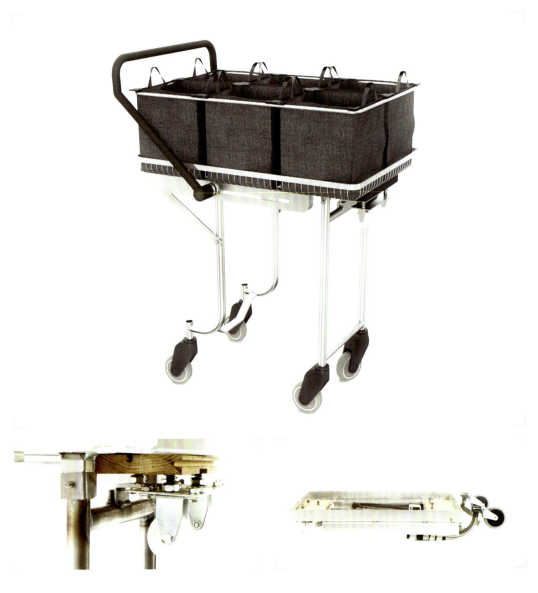

Supermarket shopping is a lengthy and challenging task that many reluctantly endure weekly. Combined with badly maintained and unhygienic supermarket trolleys, shopping can become an unpleasant and tedious task. This personal shopping trolley has been designed to load directly into the back of a car boot, requiring only one push from the user. The collapsing mechanism has been fully automated to remove human error from the system, using the initial contact between the trolley and the car as a trigger mechanism. The process can be reversed allowing users to unload the trolley and wheel it directly into the house, then unload it by removing each of the basket's six canvas bags. These canvas containers remove the need for environmentally unfriendly plastic bags, turning a bag for life into a trolley for life.

Philip Thomas
Industrial Design and Technology

70% of domestic fridges across Europe are kept at temperatures above the recommended 5°C. This can compromise the quality of the food being stored and can lead to the growth and spread of harmful bacteria. Monitoring the fridge temperature will enable users to take control of storing their food safely at the correct temperature, and reduce the risk of food poisoning. In collaboration with LCR Hallcrest, a product was produced to act as a visual warning system, changing colour if the temperature of the fridge is outside of the appropriate range of 0°C to 5°C, while displaying an accurate temperature reading using thermochromic materials. It makes monitoring the fridge temperature easier and more understandable. The product has been designed so that it is convenient to position within the fridge whilst retaining visibility.

Madeleine Carver
Product Design

Stable Temperature Lunchbox

Fresher food for longer

The Stable Temperature Lunchbox addresses the problem of keeping food cool whilst travelling to work or on days out. A cooling constituent is used to keep the contents cool for several hours. This function allows the food to be kept at a stable temperature of 4 - 6°C for several hours, resulting in the food being fresher, of higher quality and also in reduced food spoilage and possible illness from bacterial growth. In addition the Stable Temperature Lunchbox also has other design features that are unique and offers the consumer added value from the product. It utilises material choice and manufacturing methods to be as environmentally efficient as possible, whilst being intended for a mass market.

Mark Connor

Product Design

With the extensive range of products available, the design of the kitchen sieve has remained the same despite having numerous problems, including being difficult to clean and uncomfortable to hold. Nodo consists of a frame with two independent wires threaded through, perpendicular to each other, in order to form the mesh. The frame features a built-in mechanism that allows an easy shearing motion to be made, causing the wires to slide over each other, removing any trapped food from the mesh. The form of the sieve is designed with both use and storage in mind. The design takes up little space in comparison to current kitchen sieves. The user can comfortably hold the sieve and operate the mechanism all from one grip position, allowing quick and easy cleaning.

Tom Jay

Product Design

The Ultimate Luggage
No need for check-in

This collaboration with PBFA (Peter Black Footwear & Accessories), this brief seeks to deliver 'The Ultimate Luggage Design' that will meet the needs of the world's most demanding business people. As time is a crucial factor for the frequent business traveller, it is important that they remain stress-free and experience a smooth easy journey throughout their trip. This innovative expanding concept of the "F86" is like no other luggage on the current market. With existing cases on the market only having the ability to increase of up to 25% of its original size, the F86 can achieve a staggering 86% increase, beating all of its competitors. With the closed and expanded state of the case, it is able to fit airline hand baggage allowance allowing the user to keep their baggage with them at all times without the need for checking-in.

Michael Chung
Industrial Design and Technology

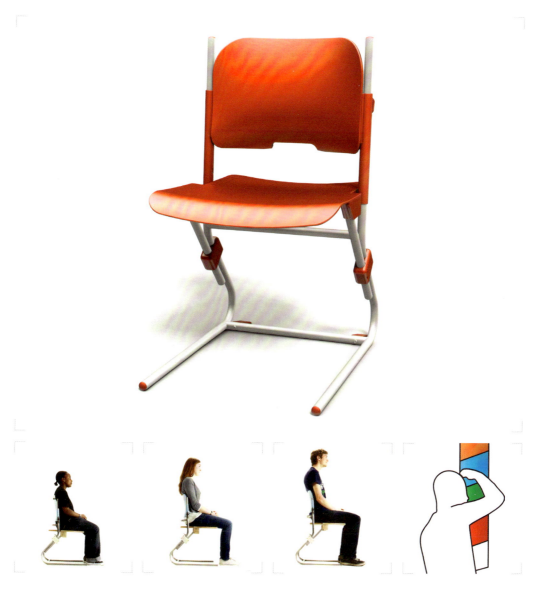

Existing school chairs do not fit the students using them. Students grow at different speeds, and the current 'one size fits all' approach results in an uncomfortable learning environment with negative health impacts. In collaboration with the expert ergonomists at FIRA International, this project aimed to design a secondary school chair that adjusts to allow a better fit between the students and the seats in which they learn for long periods each day. A solution was developed that allows the student to quickly adjust the chair by moving the key dimensions in one smooth motion. A height chart is used to find the correct setting. This allows a comfortable seated position for many different sized students. The chair cannot be adjusted once seated, preventing distractions from lessons and encouraging a positive learning environment.

Adrian Hodder
Product Design

DSLR Motion Control Rig

Time-lapse photography with a changing field of view

Time-lapse video is typically shot by using a DSLR camera to periodically capture individual still frames. After all frames are captured, they are compiled into a video file, which shows a long period of time elapsing in a short clip. Conventional time-lapse videos have a static field of view. This is because it is near-impossible to move the camera slowly and accurately enough between each frame, with a typical panning shoot requiring movements of just 0.1° or less. The Autolapse System allows the producer to create time-lapse video with a changing field of view, through the use of minute and precise movements between shots. The system includes three modular movement elements, making it possible to pan, tilt, and/or dolly. The system also integrates this movement with automatic camera triggering.

Dave Anderson

Product Design Engineering

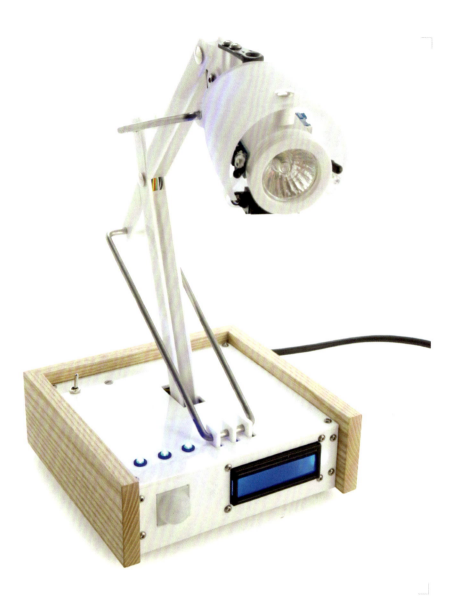

Cyclops is a motorised task luminaire that uses Kinetic Control to define its position. A number of sensors on the head and base record input from the environment and user to allow the luminaire to position itself. Head position and brightness are controlled by hand movements; the head can follow the user's hand and the light source dims up and down depending on the proximity of the users hand from the lamp head. Kinetic Control offers a novel method of interaction, increasing ease of use for many different audiences.

Alexander Ambridge
Product Design

Future Concepts for Adobe?

Adobe's continuous efforts to perfect their products on the market begs the question: what if they expanded outside of the "software-only" model and into a new deep market with the same core values? Adobe's main value is creativity so the vision for Adobe is for them to become a first party developer of hardware that will allow users to express their creativity for the world to see. This is seen in the use of high-end technology and futuristic conceptual forms, which have expanded from the current generation of devices.

The Creative Interface Suite

Future concept for Adobe

The Adobe CIS has been designed to be used by working professionals in the design industry. The product incorporates a touch screen tablet which has the capabilities of all the software products in the range including new 3D CAD software. The interface allows the user to trace, sketch or even render, mimicking the activities traditionally done on paper. The projector feature gives the user a second visual display which can be used for presentations. Alongside the touch screen interface is a holographic tablet which allows the user or clients to get closer to the design in a completely new dimension. It can be used as a show piece in client meetings to sell the idea or as a tool to increase productivity and efficiency in the workplace.

Jason Cham

Industrial Design and Technology

Adobe has always been perceived as a software only company, with their successful commercial products dominating the creative side of the software world. However, it is time for Adobe to move into the hardware market and produce signature products that run exclusive Adobe packages at optimal performance for professionals. The concept behind the dual display notebook is to change the way a standard A4 notebook works and revolutionise the art of note taking in a digital world. The Adobe notebook is a 7" dual display with a thin side profile of only 8mm. This sleek and elegant device will change the way that creativity is expressed and with the addition of a pico-projector it will also change the way creativity is shared.

Kourosh Atefipour
Product Design Engineering

Digital Art Palette
Future concept for Adobe

In their 25 years existence Adobe have revolutionised how the world engages with ideas and information. They pride themselves on delivering high quality digital experiences across a range of devices. By taking the older generation into consideration and how they are falling behind with today's sophisticated technology, the Digital Art Palette allows Adobe to bring the traditional practice of painting into the contemporary digital world by providing a new platform. This is also suitable for any user group. This is achieved using two complementary components: a digital art palette and a stylus. Alongside these two, a docking stand is included to support your new hardware wherever you go.

Michael Chung
Industrial Design and Technology

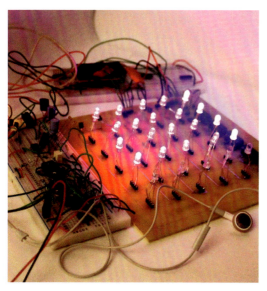

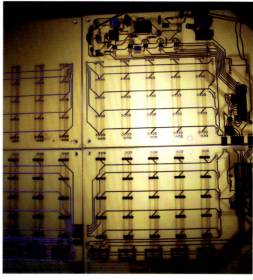

This project takes movement and converts it into light, colours and sounds. This has been achieved using a handheld device containing an accelerometer which measures the movement of the user's hand. The data is then transmitted to a wireless receiver which uses it to change the colour and lights on the 100 RGB LED array. The data received is also used to change the values of several digital potentiometers which alters the sound output on a simple music synthesiser. Overall the system incorporates 3 micro-controllers, 5 digital potentiometers, 2 wireless transceivers and 110 LEDs.

Tom Maltby
Product Design

Name Me
A lamp with personality

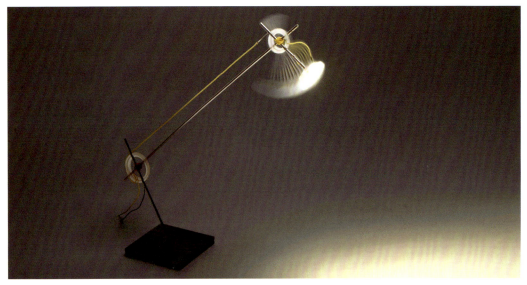

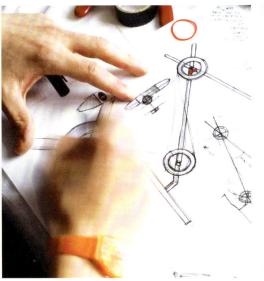

Name Me is about holistic design. Holistic design is the process of combining emotion, technology and logic into a product. In doing so, these factors are re-contextualised into a new form. Consequently, Name Me has been created around play, emotion and design. Empathy and story are also relevant parameters. The result is a desk lamp with its own distinct character and the ability to interact with its user on a playful and emotional level. Name Me is the lamp that is happy to see you.

Roman Luyken
Industrial Design and Technology

The trend towards a world where ubiquitous computing, in which information processing has been thoroughly integrated into everyday objects and activities, holds many chances and problems. Increasingly smaller computers and chips will make our lives more convenient and interconnected. Creating a platform where the advantages of low voltage power could be exploited was the starting point for this project.

The developed track adaptor system enables the elimination of power supply units and excess cable and enables quick and flexible setup changes of a wide range of devices. It offers a discrete, uncomplicated and easily installed power distribution for changing room functions and layouts in environments such as exhibitions, pop-up shops and galleries.

Emanuel Köchert

Industrial Design and Technology

D-Rise
A drum set storage solution for small spaces

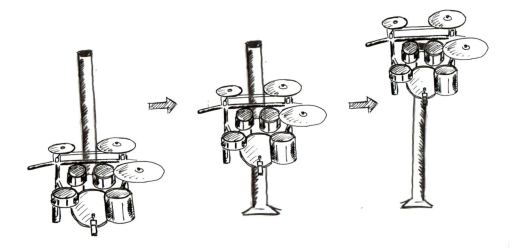

The average drum set occupies at least 15 sq. foot of space and as rapid urbanisation continues, every inch of space will need to be efficiently utilised. Drummers struggle to accommodate a full size acoustic drum set - the future rock stars of tomorrow need a solution. Each component of the drum set attaches to a customised drum rack, which fastens onto a wall mounted frame.

The drum set then simply slides up towards the ceiling, effectively utilising the upper dead space in the room. The drum rack can be detached from the frame with the set attached and rolled to the desired location. Drummers will no longer have to live in cluttered spaces or dismantle and hide their set when not in use. They can simply slide it up and proudly display their drum sets.

Rohan Malhotra
Integrated Product Design

A flautist will warm their flute to playing temperature then make any necessary adjustments so that it is in tune. However, if the temperature of the flute fluctuates, the pitch will be affected. A musician would be unaware if this happens, and will not know if they are still in tune until they play. This lack of feedback between a flautist and their instrument can lead to embarrassing tuning errors in a concert.

The discreet, thermochromic temperature indicators clip onto the flute, allowing the musician to be better aware and make temperature adjustments to help keep their flute in tune. They are quick to evaluate and act as a reminder to warm the flute before playing. Also, they would be a beneficial way to educate novice and amateur flautists of temperature effects.

Emma Tuttlebury
Product Design

Interactive Ukulele

Play chords on a ukulele using a computer

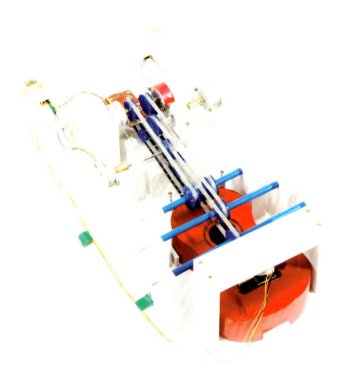

A technical experiment to see if it is possible to mimic the mechanics needed to play the ukulele. This prototype is designed as an interactive product where the user can use stepper motors controlled by a computers to try and play a chord. Most self playing guitars cheat by having individual actuators that stamp out the chord shape and therefore only play a preprogrammed set of chords. This is limited in terms of use, very cost ineffective and has very little actual user interaction. This self playing ukulele uses 4 stepper motors rigged to pulleys, meaning that potentially any chord shape is achievable, even ones not humanly possible. The addition of a point and click interface means also that the user can choose different chords and the computer will calculate the quickest route to the next chord from its current position.

Victor Jeganathan

Product Design

Electrostatic drivers are used by some of the world's best loudspeakers, but come with a premium price tag. This project looks into the feasibility of simplifying the manufacturing processes through design, to reduce the cost for the end user. Retaining the quality sound capabilities of electrostatic drivers whilst reducing the price, the exotic driver is in direct competition with more affordable and well known home theatre and hi-fi brands on the market today. Gauss is a cost effective solution to high quality audio for the average user.

Allan Lowther
Industrial Design and Technology

myLab
Drug purity testing device

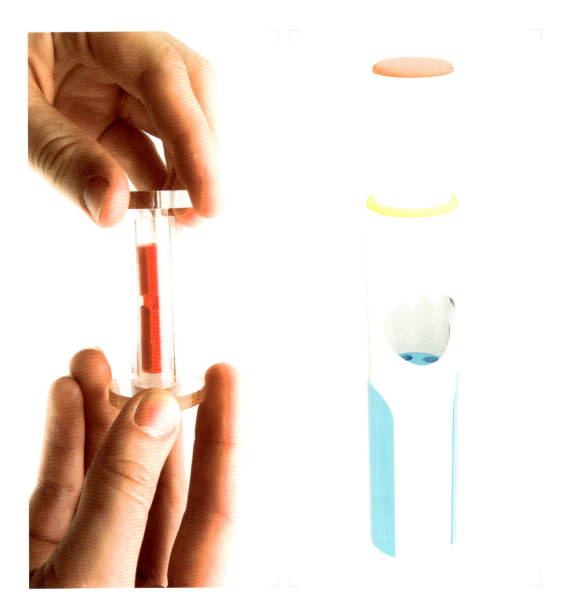

Drug users will experiment and addicts will abuse substances whether they are legal or not. People take drugs and drink alcohol for a variety of reasons including peer pressure, escapism, or recreation. Unlike alcohol however, drugs come without any indication of purity, strength or ingredients, which can increase the likelihood of long term health problems, or in some cases death. This is a particular issue with ecstasy pills which contain a variety of different ingredients, impossible to determine for the average drug user leading to unnecessary risk. This product will reduce risk to vulnerable users by providing a portable, one use only, drug testing device. In this way individuals will develop an awareness of the substances they are taking and attention will be drawn to an issue which needs to be tackled before it is too late.

Hannah Devoy
Industrial Design and Technology

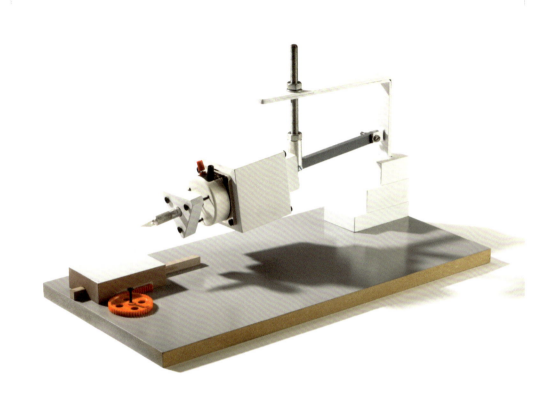

This project looks at creating a dynamic scalpel blade able to cut through material that closely relates to human tissue. The application of this product would be aimed at ophthalmic surgery, in particular cataract corrective surgery. In order to achieve this, a series of experimental rigs were manufactured to test a range of different frequencies and amplitudes. The results of these tests were examined and a frequency and amplitude were selected that obtained the most suitable cutting pattern for the application.

Richard Harris
Industrial Design and Technology

Swinging Arm Testing Equipment

Human arm simulator for impact-testing protective equipment

The aim was to create a piece of equipment that could test the effectiveness of pieces of personal protective equipment against violent attacks. The brief was set by the Firearms and Protective Equipment program, a part of the Home Office Scientific Development Branch. This division is responsible for providing technical support to all UK Law Enforcement Agencies, which includes the testing of equipment used by members of these agencies operationally. The primary design focus was the operational threats faced by officers in the UK Police Forces ranging from Routine Patrol Officers to Public Order Officers. The resulting product is a pneumatically driven swinging arm that effectively replicates the biomechanics of a human arm capable of wielding a variety of weapons in a number of different striking motions.

Dan Wilkin

Industrial Design and Technology

After reviewing their current vandal resistant panels, a package of ideas has been developed for Door Entry Direct Ltd. Currently the company has over 400 combinations across 7 different panel sizes, each one separately produced and stored, with increased production time, costs and unnecessary stock holding. Firstly, providing functional and aesthetic changes to their current panel ranges to be incorporated using current production techniques. Secondly, exploring Touch Through Metal technologies as a potential way to improve the panels' vandal resistance whilst lowering costs, stock levels and production times. Thirdly, creating a new modular panel within the constraints provided. The design is a universal frame and connection system which can be adapted to house a vast array of different modules from speaker to RFID.

Jonathon Rugg
Industrial Design and Technology

Medical Disinfectant Dispenser

Improving hand hygiene behaviour in hospitals

Nosocomial (Hospital-Acquired) Infections are an issue medical facilities face worldwide, affecting 10% of all occupied or treated patients and resulting in over 135,000 deaths annually in Europe alone. The Medical Disinfectant Dispenser monitors the hand hygiene behaviour of visitors and staff to ensure that the transmission of bacteria to patients is kept to a minimum. In order for a visitor to gain access to patient rooms, the user is required to disinfect hands at the allocated dispensers. Once the hands have been disinfected, the user will place an iButton, previously received at the reception, to the 1-Wire device integrated in the Dispenser. The visitor can now use the iButton to access the patient's room for a set period of time, before hands will need to be re-disinfected upon re-entering the room.

Philip Zeitler

Industrial Design and Technology

Cataract treatment is an extremely delicate procedure that requires precision and accuracy. Current methods are deemed very difficult due to the high level of skill needed by the surgeon and as a result the advanced surgical scalpel has been thoroughly researched and developed. The Advanced Ultrasonic Scalpel is set to increase the procedure's success rate and to simplify the procedure for surgeons by introducing ultrasonic vibration to the blade head. The project has seen many developmental phases including testing and prototyping to find the optimal variables for successful use. Research identified the correct ultrasonic vibrator for the application. The application of this project is not just limited to the human eye, but also to other high collagen areas.

Kourosh Atefipour
Product Design Engineering

A Disposable Exhalation Block and Silencer
Improving the hygiene of the SLE 5000 neonatal ventilation unit

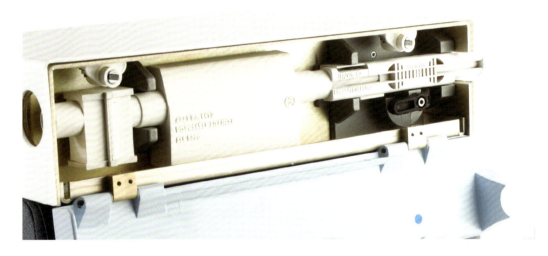

This is a collaboration project between SLE and Brunel University. The brief was to improve the hygiene of the neonatal ventilation unit the SLE 5000 by redesigning two of the components as one-off disposables rather than autoclaveable products. This included a new exhalation block, a newly redesigned silencer and later on led to developing a new bacterial filter that could fit inside the SLE machine.

Victor Jeganathan
Product Design

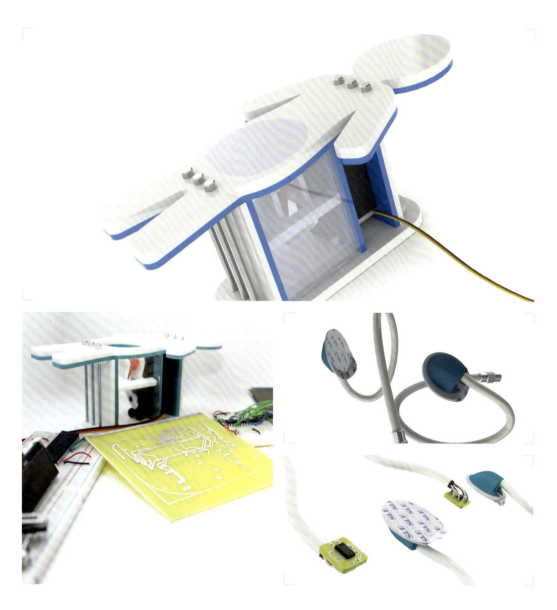

This was a collaborative project with SLE Limited, who wanted a new way of detecting the breathing cycle of neonatal infants so that the ventilators could be better synchronised with the patient. This is achieved by measuring the movement of the abdomen. The project focused on developing a sensor system using an accelerometer to measure the movement of the abdomen. A test rig was also developed to test the designed sensor system. The response time of the accelerometer needed to reduce the time lag to less than 25ms as a greater lag can cause the neonate distress. The sensor system also needed to distinguish between different breathing events such as coughing, sneezing and crying. The final project consisted of a test rig and visual display of the movement, with a separate module which could be integrated straight into a neonatal ventilator.

Tom Maltby
Product Design

Aura

Preventing infant scalding using temperature sensitive flow control

Scalding is a serious problem when considering those most vulnerable such as the aged, infirm, or infants. Each year over 300 serious scalding incidences occur involving infants and their feeding bottles. Aura completely prevents the flow of hot liquid to the infant when the liquid is in excess of 37°C, using a thermo-responsive mechanism integrated within the bottle. The strong potential for adaptation into similarly vulnerable markets has guided the development of the mechanism, whilst the precise temperature operating range has dictated specific component configuration. Aura is an entirely unique product that satisfies a broad market need, providing reassurance and confidence in infant safety that is especially sought after by first time parents.

Jade Boggia
Product Design Engineering

Pointe shoes are made from pasted layers of paper, cardboard and satin with a sole and shank of leather. Traditional pointe shoes are used by all levels of dancers and are essential to ballet as we know it. When used, pointe shoes absorb perspiration, which is essential to the process of them moulding to the dancer's feet. Over time this weakens them. This leads to expense, unnecessary replacement and possibly injury. The pointe shoe dryer extends shoe life by drawing moisture away. This can be done many times before regeneration through heating.

Rebeckah Rose

Industrial Design and Technology

Temperature Sensory Sleeve
Important temperature information for athletes

The Temperature Sensory Sleeve helps prevent injuries sustained during sporting activity, indicating when an athlete has reached their optimum body temperature needed for sports using thermochromic materials. Up to 19% of all acute injuries seen in emergency rooms are sports injuries with football injuries accounting for 23% of all injuries. Many of these injuries can be prevented through the correct process of warm-up and pre-match exercise. This product also acts as a learning aid for aspiring young athletes in terms of correct warm-up methods.

Emmanuel Hope
Product Design

The Jump Shot is the most commonly used in basketball. The objective was to improve player's efficiency and shooting ability during their training period. Often a player's shooting accuracy decreases as time progresses, due to player fatigue and a decreased endurance and shooting strength. The Shooting Aid aims to work as a circuit-training device that records the player's ability to make a series of shots. Once each shot is made within the allocated time a positive sound will go off, stating that the shot was made within the time limit, but if the average is not met a negative alert sounds. The player can train to beat their own personal best as well as compete with teammates. The product will have three stages to cater for each player's skill level; the better the player, the higher the level.

Seun Babatola

Industrial Design and Technology

High Altitude Oxygen Mask

Coping with high altitude mountain climbing

In the world of high altitude mountain climbing, there has been a great deal of technological development. There remains one aspect, somewhat surprisingly, that has not progressed at the same level - the oxygen mask itself. The masks in current use are essentially just adapted aviation systems which are intended for use in a completely different environment and most notably within the protection of a cockpit. When these systems are then thrown into use in such extreme climates present at the top of Everest and other 8,000m+ mountains, some real issues begin to emerge. This project involved evaluation of these issues and development of a mask that is better suited to use in the environment within which it will be used.

Michael Day

Industrial Design and Technology

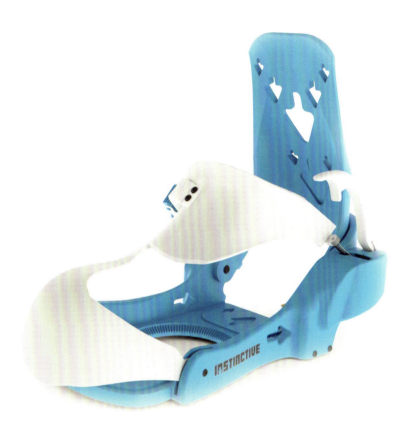

Instinctive Bindings is a completely automated snowboard binding system for freestyle snowboarders. This binding design allows the user to enter the binding hands-free yet still retain the control offered by manual strap bindings. Its robust form and mechanism has been specifically designed to withstand high impact and be instinctive to use. Details in the form of colours and shapes subconsciously provide information to the user of how to operate the bindings. Once the user enters the binding they trigger the mechanism that clamps the uni-strap shut. To exit, the clip is squeezed and the handle is pulled.

Matthew Rowinski
Industrial Design and Technology

Puncture-Proof Bicycle Tyre

Never worry about a puncture again

This product is a solution to the global problem of bicycle tyre punctures. It is primarily aimed at commuters or cyclists who lack the knowledge or time to change or repair tyres. The radical spoke design on the product absorbs harsh impact in conjunction with providing essential structural strength, supporting both the bicycle and cyclist. The closed cell polyurethane and the use of D3O help the tyre behave like a pneumatic tyre. Using polyurethane allows for cheaper manufacturing and effectively cheaper tyres. The hassle of carrying bicycle pumps is eliminated as are puncture repair kits and any other of the tools required with pneumatic tyres. This concept eliminates the need for maintenance and in today's world, time is money. Registered Design.

Jason Cham
Product Design

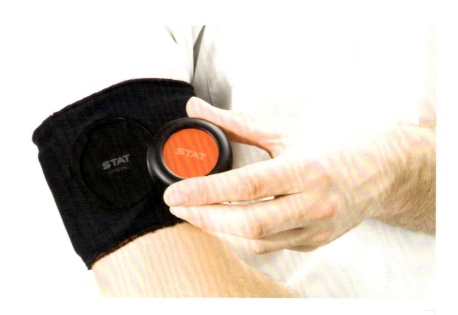

Sports Team Analysis and Tracking (STAT) is a wearable tracking device that allows outdoor football games to be recorded and analysed. Each player can replay their movements and gain statistics on how they performed alongside their teammates. The accuracy of GPS technology has improved to capture positions within a few metres at a low cost, allowing teams at all levels of football to generate player statistics.

The STAT armband is designed to seamlessly fit into a player's match day routine. A key feature gives the players the chance to highlight moments during the match. A tool to aid player development, but also to enhance the enjoyment of taking part in sports; users can revisit the game with friends, share incidents, track improvements and compare themselves to the professionals.

Mike Puttock

Industrial Design and Technology

Patent Reinvention

A research study into intellectual property strategies for design managers

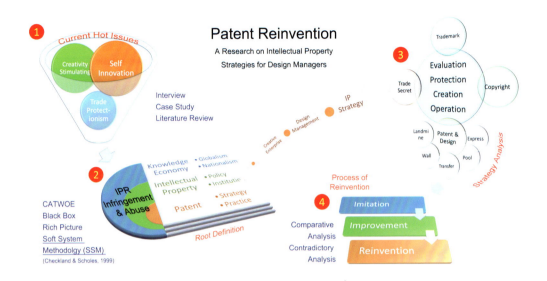

Patent Reinvention

A Research on Intellectual Property Strategies for Design Managers

Infringement is a cancer, which destroys healthy economic systems and discourages the growth of independent innovations. How to exterminate patent infringements has become a vital issue in global trade. The aim of this research is to provide an integrated patent reinvention toolkit for design managers and intellectual property (IP) strategists in order to exterminate infringements. Lord Clement-Jones said: "The UK encourages creative industries and welcomes mutually beneficial IP co-operations with Russia, China and other countries." At the beginning of 2011, the British government carried out a patent box project to stimulate R&D. Compared with developed countries, China is facing IP threats. In 2010, 1/3 of the 337 cases happened in China, all of which were patent infringement prosecutions. Patent reinvention, as a method of patent transfer, is an economic IP strategy to gain independent IP rights and avoid infringements. Reinvention is a part of the inspiration of innovation. US President Obama also emphasised 'reinvent' a lot in the State of the Union of 2011. This research consists of five steps: background and context

investigation → IP strategies investigation → typical invention process analysis → a patent reinvention toolkit → evaluations. IP strategy is a worldwide and complex problem requiring soft system methodology (SSM) including methods of CATWOE, Black Box and Rich Pictures in order to find out the root definition and narrow down the scope of the topic. Literature review and interview are also the main methods. Besides, this research adopts comparative analysis and contradictory analysis to study patents and infringement cases. Infringement is becoming a world IP game. To punish infringements is one thing. However, trade protectionism is another thing which is a zero-sum game. Infringement investigations aim at market access bans which stop rival's activities within its IP lifecycle. According to case studies such as LG refrigerators and GFCI, an infringement investigation prosecution requires on average $1-6 million in legal costs and takes 1-6 years. Customer losses and reputation damage are immeasurable. Thus more infringement investigations stand at the opposite side of inventors and IPR now is abused by some strategists on purpose. This phenomenon is a reflection of the nature of IP as a state-based duty-bearing monopoly. Somehow, monopoly is rising while duty is lost. To defend the inventor's dignity or let money eat your conscience? That is a question to design managers and IP strategists.

"UK encourages creative industries and welcomes mutually beneficial IP co-operations." Lord Clement-Jones

Zheng Li
Design Strategy and Innovation

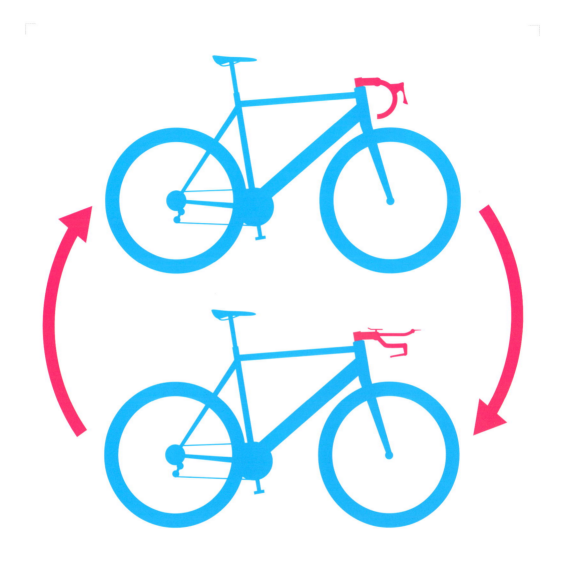

Many people compete in both Road Racing and Time Trial disciplines but can only afford to own one bicycle and this can affect their ability to compete in both disciplines effectively. Altering a Road cockpit layout with add-ons is a common compromise, but there are disadvantages to this, such as increased drag. The only other option would be to switch to a Time Trial cockpit, therefore having to re-route all of the gear/brake cables and re-configure the gears and brakes afterwards - this is not desirable. The aim of this project is to provide a solution for cyclists that enables them to reap the benefits of both types of cockpits, while only possessing one bicycle. This will give them the freedom to compete effectively in both Road Races and Time Trials with no disadvantage.

Max Woźniak
Product Design

An Easy to Assemble Rigger Attachment

Easing the process of attaching riggers to rowing boats

The current method of assembling a rigger to a rowing boat can be a very tedious and time consuming task. If incorrectly assembled these riggers must be disassembled and then reassembled correctly. This can become a logistical nightmare when managing a large group of juniors and when a large number of boats and riggers are involved. The final design is a sliding mechanism that is retro-fitted onto to an existing boat and rigger, allowing it to be easily manufactured and sold to the rowing community. The product aims to simplify the rigging process; eliminating the need to screw nuts and washers and reducing the time it takes to check the correct assembly of riggers.

James Clarke
Industrial Design and Technology

With its many health and environmental benefits, growth in the popularity of cycling has many positive effects. SPRUNG is an automatic bicycle transmission with an expanding sprocket which provides the gear range of a three speed bicycle. SPRUNG removes the hassle of gear control by the user. The individually sprung sections are moved through a positioning guide once every rotation which moves the section into the correct gear position. The guide is altered using a centrifugal governor which translates the bicycle wheel's rotational movement into a linear push and pull on the gear cable. Designed for leisure, SPRUNG aims to encourage more of the population to take up cycling.

Tom Reader
Product Design

Reinventing the Wheel

Real-time bicycle tyre pressure regulator

The ideal pressure for a bicycle tyre varies considerably depending on the surface involved. On a smooth road, hard tyres are desirable to minimise rolling resistance. In muddy conditions much softer tyres provide better traction and ride comfort. This product enables cyclists to adjust their tyre pressure with the bicycle in motion. The system uses a refillable aluminium CO_2 cylinder mounted on the bicycle frame to deliver pressurised gas to the rear tyre. A link is made to the rotating wheel using a specially modified hub containing rotary seals. The pressure is adjusted via a small lever on the handlebar which provides an input to a microprocessor. This microprocessor can then control the operation of two solenoid valves to increase or decrease the tyre pressure.

Ross Thompson
Product Design Engineering

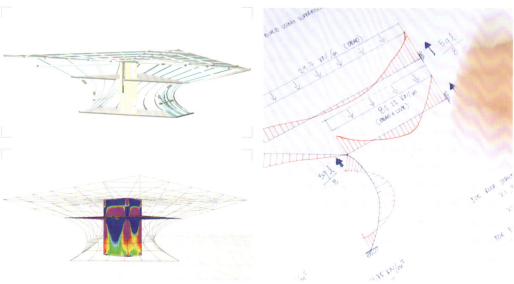

The Twist Exhibition Centre was designed to project a modern image in Hyde Park. The result is a two storey exhibition centre that accommodates all functions with a recreational area, exhibition rooms and a restaurant. From a distance the building looks like a series of twisted columns, an impression created by the "turning effect" of the circular pipe columns. The elements support each other and combine to form the complex geometry of the building. The overall steel structure of the façade and roof is welded firmly to form a single entity. There are no movement joints between the individual members in the steel frame. Furthermore, a concrete share core is located in the centre of the structure to provide stability and access to services. The materials used in the building are steel and glass predominantly.

Abraham Garcia
Civil Engineering with Sustainability

The project involved the design and manufacture of an electric superbike. The donor bike, a 2009 Yamaha YZF R6, was adapted to become the first road legal prototype bike developed by Brunel. The project involved all areas of mechanical engineering and design including, but not limited to, mechanical and electrical part design, simulation, manufacturing and testing. The project involved investigation and design of a unique KERS package and a two stage performance switch. The final outcome of the project was a working prototype of the design, which is ready for extensive road testing and development. Alongside fulfilling the design brief, the project also aimed to allow students in the next few years to continue the development of the bike, enabling Brunel to further its knowledge of electrically powered vehicles.

D. Millard, C. Youell, T. Rutter, T. Ellson
Mechanical Engineering with Automotive Design

The principal objectives of the project were to design, manufacture, build and race the BX-11 at the 2011 TT Zero event held during the famous Isle of Man TT. The TT Zero, an event for zero emissions vehicles is now in its second year and is attracting competitors from around the world and will be broadcast globally. The team have focused the design upon a Triumph Daytona 675 donor chassis and redesigned the power and drive train, structures and control systems. Models were also developed to simulate various configurations of an electric motorcycle in order to adapt, dependent on their operational environment. This was based upon minimising the lap time of various tracks including the TT Mountain Course.

E. Bright, C. Kneller, A. White, T. Clarke
MEng, Motorsport Engineering, MEng with Automotive Design

Motorsport Airflow
Ventilated racing helmet

This project is aimed to provide the racing driver with sufficient ventilation to enhance their performance physically and mentally. The idea originated from track events where companies use inexpensive helmets and the intended concept will aim to reduce the heat transfer into the padding. The helmet is structured to allow free air flow as well as withstand impacts, through the reduction of padding. Different structures that provide ventilation, such as bicycle helmets, were taken into consideration to combine speed and safety in a light-weight helmet.

Nathan Chan
Product Design

Modern vehicles keep the driver's attention within the interior environment, making them isolated from the outside world, and less aware of the activities occurring around them. While current parking aids and high tech systems are designed and applied with good intentions, the situation and visual awareness of the driver has been compromised. Although these lapses in attention may only occur for miniscule amounts of time, they are long enough for an accident to occur. Angel, aims to form a parking aid without compromising situation awareness, making for a safer, more reliable source of support for the driver. It provides an appropriate solution for the driver to protect their valued assets (alloy wheels, vehicle, etc.) from accidental damage during parking/manoeuvring and reduce the risk of a collision with an inanimate object.

Jayson Tulloch
Product Design

Brunel Racing BR12

Class 1 Formula SAE Car

Brunel Racing is a team of undergraduate engineering students that design and build a formula-style racing car to participate in the Formula Student competitions, attracting over 100 universities in the UK and Germany. Brunel Racing cars have won many awards, including honours for engineering excellence, most desirable engineering product, best engine and most fuel efficiency. In 2009, Brunel Racing competed against over 80 teams at the Silverstone Circuit and the Hockenheimring in Germany, finishing as the second highest placed UK team at both events.

BR12 Design Overview

BR12 is the team's 12th consecutive entry into the Formula SAE Class 1, with great emphasis placed on reliability, increased power and driveability. Consideration was given to improving the environmental impact of the car through the use of recyclable materials and optimised design. For example, the previous carbon fibre steering wheel has been replaced with a recyclable vacuum formed plastic design. The ergonomics have also been enhanced, with improved location of suspension components for ease of adjustability and a multifunction display to allow the driver to monitor the car and extract the highest possible level of performance.

BR12 Specifications

Engine Type	Yamaha R6 – 599cc
Max Power	94 bhp/ 70 kW
Top Speed	118 km/h
Weight	200 kg
Chassis Weight	35 kg
Length/Width/Height	2406 / 1475 / 980 mm
Wheelbase	1580 mm
Track front / rear	1200 / 1175 mm
Drive Type	Chain & sprocket
Chassis Construction	Aluminium monocoque/steel spaceframe hybrid
Honeycomb Material	25mm thick / 0.9mm 6061T6 skins
Spaceframe Material	25.4mm OD 4130 steel tubing

Fuelling	Common rail injection, Tata Nano fuel pump	

BR12 Team

James Spencer   07825266040
                     spannnner@hotmail.com

Martin Newman   07738 490896
                     martinpnewman@gmail.com

Rob Murton   07759 942442
                     robmurton@hotmail.com

Matt Carey   07989 740 508
                     matt.carey88@gmail.com

Neil Whitehouse   07905 676699
                     neil.whitehouse1@gmail.com

Tom Cakebread   07763639494
                     tomcakebread@hotmail.com

Donna Law

Ben Cottle

James Luetchford

Ben van den Bos

Adam Gumbrell

Dave Walker

Fuelling
Common rail injection, Tata Nano fuel pump

Differential
2010 Drexler Formula SAE limited slip differential

Suspension Type
Unequal length carbon-fibre A-arms. Pullrod actuated spring/dampers

Wheels & Tyres
13" Braid alloys, 20.5 x 7.0 - 13 R25b Hoosier slicks

Driver Size Adjustments
Individual foam seat inserts

Driver Visibility
105° visibility each side from straight ahead

Gear Actuation
Electronic semi-automatic paddle shift system

Clutch Actuation
Electronic actuation paddle on steering wheel

Instrumentation
Custom wheel /dash - multifunction display

Brunel Masters Motorsport

Formula student single seater race car

 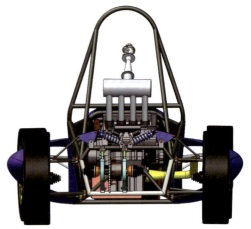

Team Profile

Brunel Masters Motorsport-07 is the seventh team representing the Brunel University in the IMechE Formula Student competition. The team consists of graduates pursuing a Masters degree in Automotive and Motorsport Engineering within Brunel University's renowned School of Engineering and Design. This year, 23 students from 5 different countries with diverse backgrounds, specialising in different disciplines, aim to produce a technically simple, running car to compete in both Class 2 Formula Student UK and Class 1 FSAE Hungary. The team follows a flat hierarchy, providing the ideal platform to innovate, obtain critical peer review and facilitate knowledge sharing.

Length / width / height	2967 / 1459 / 1305 mm 1650 mm (wheelbase)
Weight	330 kg
Suspension	SLA setup (A shaped). Front pull rod (Horizontal dampers), Rear push rod (Oblique dampers)
Wheels	202 mm wide, 2 pc Al Rim
Brakes	Cast Steel, Hub mounted (fixed type), Front (200mm), Rear (150mm)
Chassis	Triangulated alloy steel spaceframe
Engine	Electric motor for 2A Yamaha YZF-R6 four stroke in line four. 65.5mm bore. 44.5mm stroke 4 cylinder. 599 cc
Fuel	Type of Energy Storage for 2A 98 RON Unleaded
Max power/max torque	85.5 kW @11500 RPM/ 58Nm @9000 RPM

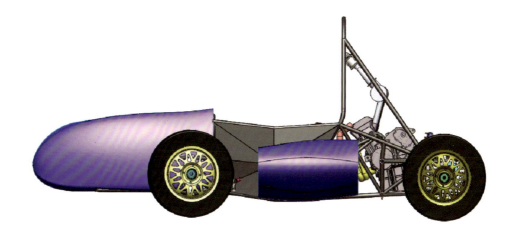

BMM- 07 Team

Ashwin Raut
raut_ashwin@yahoo.co.in 07740955908
Cillian O'Connell
lordfarquaad2nd@hotmail.com 07788813910
Ian Fulton
ian.fulton06@imperial.ac.uk 07738400529
Kunal Misra
misra.kunal@gmail.com 07404082842
Rushikesh Koli
rushikoli10@gmail.com 07740979968
Soumya Mishra
soumya-mishra@hotmail.com 07586725754
Bharath Kumar Puttur
bharathkumar122@gmail.com 07736905942
Kunal Dhande
kunaldhande@rediffmail.com 07574597968
Nithin Ambady
nithin.ambady@gmail.com 07896768967
Prasad V.G.V
vgv.prasad@gmail.com 07736904093
Sulabh Dhingra
dcemech87@gmail.com 07686725788
Aikaterini Fotoglou
aikaterini.fotoglou@gmail.com 07729836344
Chitamorn Pipatanamongkol
first_86@hotmail.com 07540167957
Iordanis Paisoglou
p.iordanis@hotmail.com 07580366511

Rahul Mishra
judopriest@gmail.com 07904240660
Sanjay SS
sanj.ss01@gmail.com 07526011197
Sanjit Plaha
sanchi.plaha@yahoo.co.uk 07960077709
Yash Gandhi
yash286@gmail.com 07791605429
Canev Civelek
ccivelek@hotmail.co.uk 07958200093
Chaiyapongse Limpanonda
climpanonda@gmail.com 07730432494
Deepak Dev
adeepakdev100@gmail.com 07845833646
Kushal Agarwal
dragooracing@gmail.com 07740896935
Vivekananda Prabhu
vicky_dj83@yahoo.co.in 07550045526

Brunel Racing Formula SAE Vehicle

Electronic instrumentation and control system for a Formula SAE vehicle

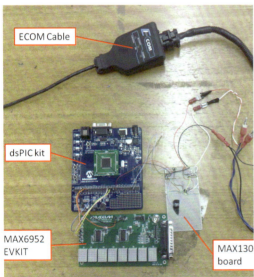

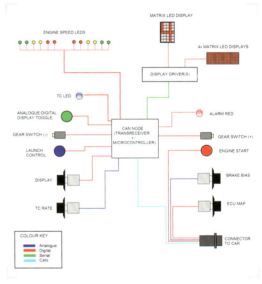

The electronic instrumentation and control system was designed for Brunel Racing's Formula SAE vehicle. It features an innovative display method for relevant information to the driver regarding the car's status and lap times. The system communicates with the car's ECU via the CAN (Controller Area Network) data bus to gather information from the multiple sensors in the car. The information is then displayed graphically on a matrix display. The great advantage of infographics is that the driver can judge the status of any system much quicker than reading a number; e.g, the fuel level will be displayed as a progress bar instead of a 7-segment display; this is more useful in a race situation as the driver does not need to think about the numbers displayed on a screen and can stay focused on the track.

Juan Diaz del Castillo

Mechanical Engineering with Automotive Design

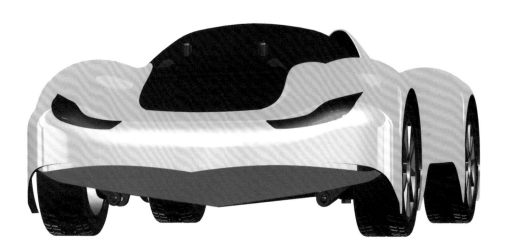

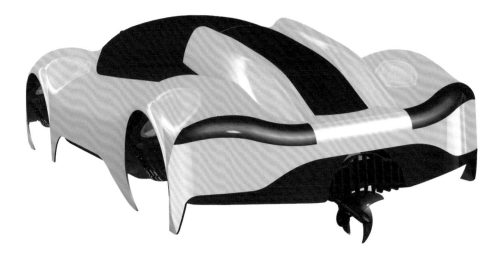

The three-channel radio-controlled Amphibious Car is four-wheel drive with double wishbone suspension systems for all four wheels. In Mode 1 - the land drive - the vehicle is configured to compete against electric motor driven on-road land based vehicles in 1/10 scale. For best performance, the car was designed with racing features such as lowered roll and gravity centre, Ackerman steering principles, adjustable steering axis geometry, ball bearings and aluminium ball differentials. In Mode 2 - water drive - the vehicle changes to amphibious drive in under 2 seconds by one button press. All wheels retract inside the bonnet and the propeller automatically shifts into position. The custom gearbox shifts 4 wheel drive mode into propeller drive and the front wheel steering automatically changes to propeller steering.

Michal Simko
Mechanical Engineering with Automotive Design

Gravity Racer Design

A competitive extreme gravity racing vehicle

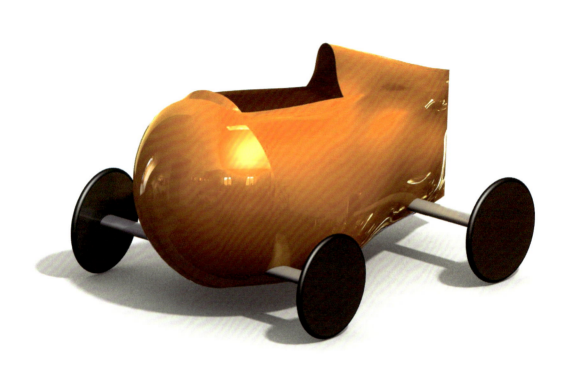

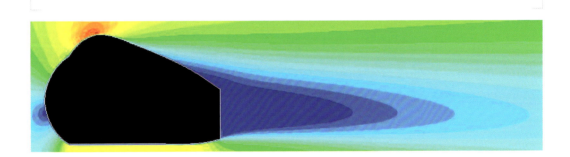

A feasibility study was conducted into whether a competitive Extreme Gravity Racing vehicle could be designed and manufactured at Brunel University. This was found to be an achievable target. A design for a monocoque chassis was selected, proposed and analysed using CFD, with an overall coefficient of drag of 1.48. The torsional stiffness of the chassis was calculated to be 12,000 Nm/deg using FEA. Other aspects of the design such as suspension, steering and wheels were considered. Their merits were reviewed and the recommended set-up includes flexible beam suspension, pitman arm steering and enclosed wheels.

Chris Knight
Mechanical Engineering

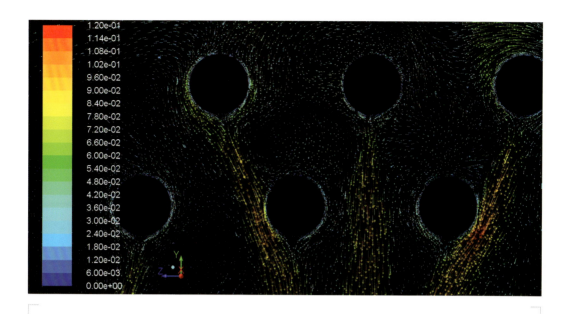

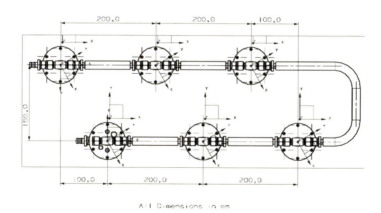

A new PCM cooling system utilising chilled water discharge was constructed and tested in a controlled environment. Internal PCM temperatures were monitored together with surface temperatures, air temperatures, and globe temperature. The addition of the modules to the space reduced the rate of temperature increase and maintained a consistent temperature within the lower and upper limits of a predefined thermal comfort zone. DSC measurements of the PCM's thermophysical properties were used to create a steady state simulation of thermal loading at peak conditions. Good validation was observed and a comparison between numerically calculated heat transfer coefficients was made with values from empirically derived equations.

Will Couch

Mechanical Engineering

Ice Accretion on a NACA 23012 Airfoil

The effects of ice accretion on the flow around an airfoil

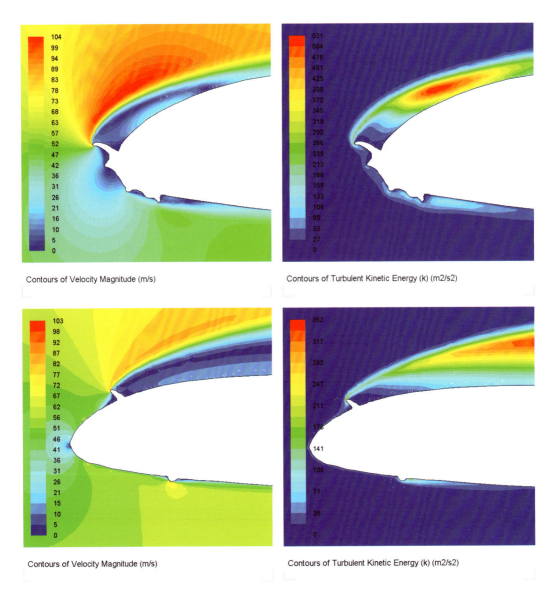

Contours of Velocity Magnitude (m/s)

Contours of Turbulent Kinetic Energy (k) (m2/s2)

Contours of Velocity Magnitude (m/s)

Contours of Turbulent Kinetic Energy (k) (m2/s2)

Ice accretion occurs when super-cooled water particles impact on a surface and freeze. This is a significant problem with aircraft as it can have an adverse effect on aerodynamic performance. This project uses numerical modelling for the analysis of three leading edge ice shapes formed on a NACA 23012 airfoil. Simulations were carried out using XFLR5 and FLUENT software packages. XFLR5 software provided inviscid results for the simple ice shapes at low angles of attack. RANS simulation using the FLUENT software gave much more accurate results for the more complex ice geometries. It provided accurate data at larger angles of attack but failed to predict stall angles. All three ice shapes showed a significant disruption to the flow when compared with a clean airfoil, which resulted in a reduction of aerodynamic performance.

George Schofield
Aerospace Engineering

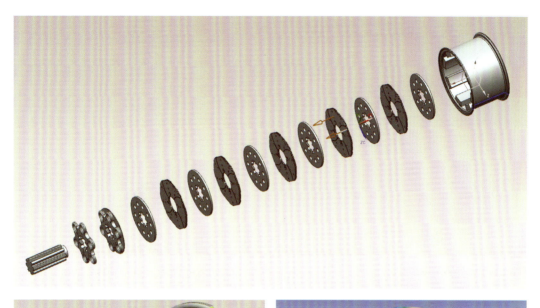

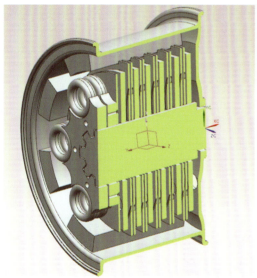

Efficient retardation is extremely important for active aircraft safety; retardation systems include brakes and hydraulic systems engines. The focus of this project was the hydraulically actuated annular disc brakes fitted to modern passenger aircrafts. The aim was to design a lighter, more efficient and airworthy braking system. The design incorporates an electromagnetic, instead of hydraulic, system. The reduction of weight coupled with newly designed silicon-carbon-matrix rotor and stator discs helps reduce the weight of the brake unit substantially and improve fuel economy. The rotor and stator discs are designed to minimize degradation and wear and reduce the level of maintenance required for inspection and repair. The design meets all relevant performance and safety criteria including international legislation.

J. Smith, M. Saeed Qureshi, F. Malik, S. Trevena

Aerospace Engineering, Mechanical Engineering

Nathan Matthews

From Pullman trains to Boeing planes

On March 20th 2011, Boeing's newest and largest aeroplane, the 747-8 Intercontinental (748 for short), took off from Paine Field in Everett on its maiden flight. Four hours and 25 minutes later, Flight RC001 landed at Boeing Field in Seattle and was declared 'ready to go fly' by the chief test pilot Capt Mark Feuerstei. In reality this flight marked the start of a gruelling test programme before the 748 is labelled airworthy, and deliveries can commence to the 748's launch customer Deutsche Lufthansa AG.

The 748 is an evolution of the iconic 747-400 'Jumbo Jet', the world's first wide-bodied aircraft, introduced by Boeing in 1970. When it goes into service early next year the 748 will be the longest commercial aircraft in history, capable of transporting 467 passengers 8000 nautical miles at speeds in excess of Mach 0.85. As well as a lengthened fuselage, the 748 features redesigned wings and the latest generation of fuel efficient engines; technology derived from Boeing's other new aircraft the super-efficient and high-tech '787 Dreamliner' (indeed, Boeing chose the '-8' designation to highlight the technological connection between the new 747 and the 787). Boeing claims that this all adds up to make the 748 the cleanest and most economic aircraft of its kind.

My own journey into the world of aviation began in 2004 as a second year student of industrial design at Brunel University. At the time, I was focused on securing an industrial placement for the third year of my four year degree. To offset the largely technical content of my BSc course, I was eager to work with a multi-disciplinary design consultancy, and I hoped to learn about the more subjective aspects of the design business. I could not have hoped for a better opportunity than that which was offered to me by PearsonLloyd. Founded, in 1997, by Luke Pearson and Tom Lloyd out of a "shared desire to

bridge the often disparate cultures of furniture and product design", the studio's portfolio spans a wide variety of disciplines including furniture, product, transport and the public realm.

I was already well aware of PearsonLloyd, not least because of their involvement in developing Virgin Atlantic's Upper Class Seat. The UCS entered service in 2003, and delivered one of the first fully flat beds in business class. Working with Virgin Atlantic to optimise their innovative 'herringbone' seat layout PearsonLloyd also applied a simple visual language, more akin to domestic furniture design, separating the seat from the functional aspects of the product. I embarked on my placement with PearsonLloyd in the summer of 2004, and quickly found myself supporting the team on a series of new products for Virgin Atlantic. These were later revealed as the airline's new economy and premium economy seats; both of which offered significant improvements in comfort and space, and shared the design language developed for the UCS. Virgin Atlantic's fleet now featured seats by PearsonLloyd across all three classes, resulting in a truly holistic interior design.

After 13 months at PearsonLloyd, I had developed a passionate interest in transport design and the airline industry in particular. Eager to pursue this as a career once I finished my degree, whilst ensuring the integrity of the projects I worked on, I approached PearsonLloyd and was delighted when in June 2006 (on the day before I graduated from Brunel) I rejoined the studio on a full time basis. Within months of being back I was once again working on projects for Virgin Atlantic, now with a more direct involvement with the client and design details.

At the same time I consciously increased my awareness of the history and culture of travel. From Pullman trains to Bannenberg yachts,

"I'm motivated by the possibilities for rekindling a sense of romance and theatre in transportation"

my insight grew and the passenger's travel experience became the focus of my interest.

In 2008 the studio got a great opportunity to build on its transport portfolio when PearsonLloyd was appointed by Lufthansa to design elements of the interior for their 747-8 Intercontinental. Lufthansa has a rich heritage in design extending across their organisation, from its iconic logo to the tableware used on board. To this day they are one of the few carriers that really invests in design, and understands the benefits it can bring to the onboard products and services it offers. For me this project marked the start of what has so far been a three-year apprenticeship in the complex field of aviation interior design.

Throughout the process we have addressed the constant commercial and regulatory challenges whilst meeting the expectations of passengers, crewmembers and maintenance personnel alike.

As the ultimate milestone of seeing the designs fly on the 748 approches, exciting opportunities lie ahead for PearsonLloyd. In an age where the act of being transported is taken for granted, I am motivated by the possibilities for rekindling a sense of romance and theatre in transportation be it on land, at sea, or, of course, in the air.

Nathan Matthews, Designer, PearsonLloyd
Graduated in 2006

Helicoptor Rotor Rig

Experimental control implementation and analysis of a scale helicoptor rig

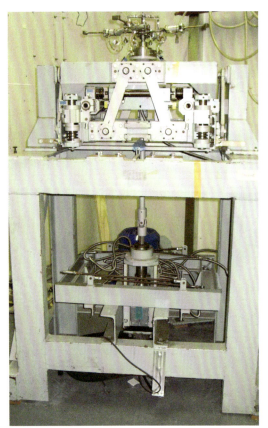
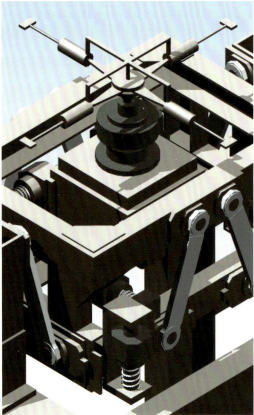

The first part of the project was to create a beams model of the Lower Rig, which could be integrated into future simulations, and perform a modal analysis. Secondly, the aim was to vastly improve upon a previous CAD model of the Upper Rig and to include missing elements and find suitable materials to simulate these.

Thirdly was to perform CFD analysis on an approximate model of the helicopter blades using a mesh deformation technique. The final major part of the project required the calibration of the strain gauges on the rotor blades, and programming of some data acquisition software.

S. Arding, N. Jackson, N. Monk, T. Thwaites

Aerospace Engineering

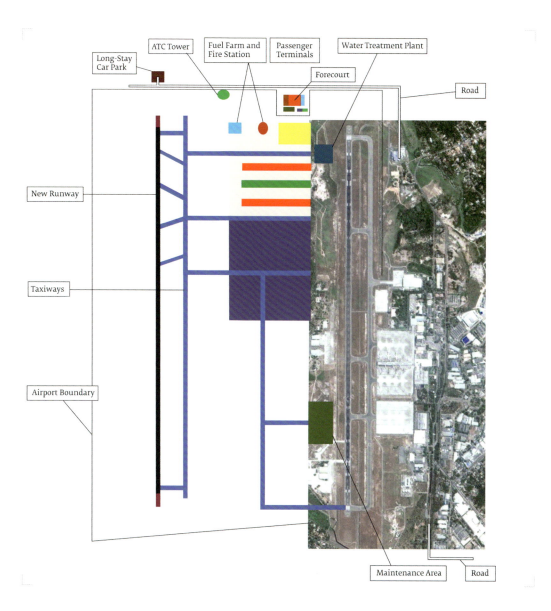

The project involved carrying out a study into the redevelopment and design of the BIA, Sri Lanka. Initially a parametric study gave an understanding of the current design and operations of the BIA. The existing airport was subjected to a SWOT analysis in order to analyse strengths and weaknesses and identify the vision of how the airport should position itself with respect to its customers and competitors.

A dynamic strategic redevelopment plan was devised. The facilities recommended by this plan were designed and dimensioned, including airside and landside facilities, and passenger terminals. The airside facilities now accommodate the A380 super jumbo with a new runway designed specifically for this aircraft. Also determined were possible environmental implications plus a cost and revenue analysis.

Jude Mudannayake
Aviation Engineering

Runway Design
Metal foam based layered composite

This new layered composite consists of airfield grade asphalt with aluminium foam and its primary application is to serve as alternative material for designing runways in remote locations. This material has tremendous compressive and impact strength to withstand the forces applied when an aircraft lands and takes off. The composite is bonded by asphalt's own adhesive quality. Using this, runways can be produced in two days, providing the subsurface layer has been compacted accordingly. The composite is ideal for building runways in any of the 26,000 Islands in the Pacific Ocean. Metal foam is a relatively new material (commercially), and is yet to be introduced into civil engineering applications. Other possible applications for the composite are surface pavements at docks and storage platforms for freight containers.

Marshel Weerakone
Civil Engineering with Sustainability

This project aimed to design and build an electrically powered, remote controlled model aircraft to compete in the 2011 BMFA heavy lift competition. The aim of the competition is to transport as much water as possible around a predetermined flight course within 10 minutes. Rules state that the aircraft must be fixed wing with a maximum wingspan of 1500mm and use only certain electrical components. Design of the aircraft is based on the study of existing aircraft along with structural and aerodynamic analysis involving FEM and CFD software. Reliance on each individual's expertise and teamwork allowed for effective integratration of different aircraft parts into the overall design. Group organisation was also critical in ensuring that the aircraft is produced on budget, on schedule and to specification.

M. Daly, J. Dhillon, R. East, M. Hamid, K. Ho

Aerospace Engineering

Container for Space Experiments

A general purpose container specially designed for use within a satellite

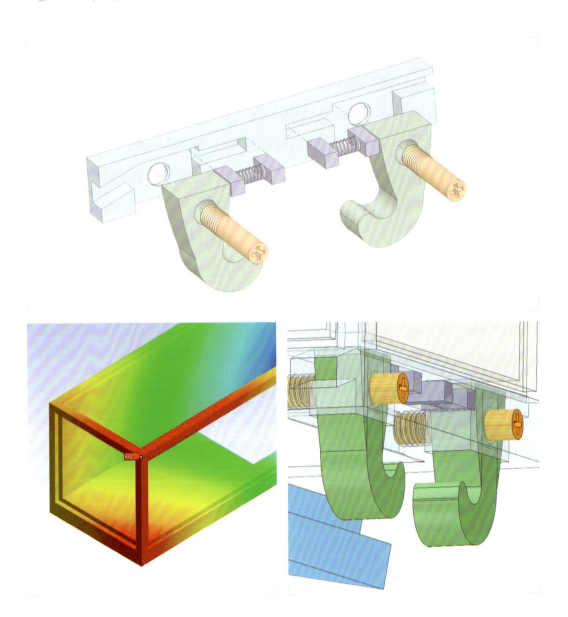

A general purpose container was specially designed for use within a satellite. the design included a mechanism to allow the container to safely dock and undock with a microgravity centrifuge. The chassis is a single piece monocoque construction made from aluminium alloy with transparent polycarbonate windows for observation. There is a trap door to access the internal volume, which is a sterilised environment containing biological material. Container transport and interaction with the internal volume is done via an electronically controlled robotic arm.

G. Dy, O. Dhillon, F. Fahimi-Rad, J. Rezaei Zadeh, R. Rezaei Zadeh

Aerospace Engineering

Carbon Nanotube Field Emitters for Space Applications

Fabricating CNT field emitters for advanced use in space applications

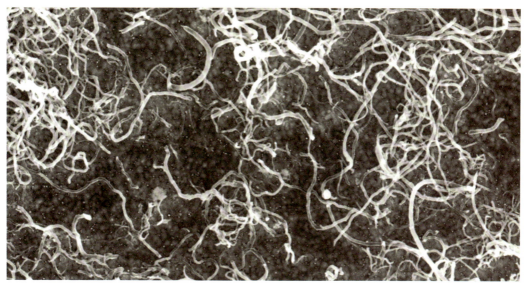

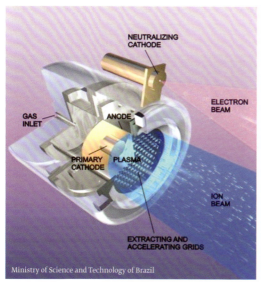

Ministry of Science and Technology of Brazil

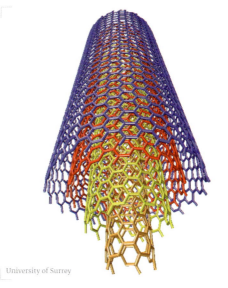

University of Surrey

Field emission is a quantum tunnelling effect under which electrons can be extracted from a material in a high electric field. Field emission cathodes are attractive electron sources for use in space because of the high current densities achievable at low mass and power. This has advantages over the most widely used electron sources based on the bulky and energy-consuming thermionic cathode based on a hot filament, in which the filament material is heated to a high temperature to release electrons. Carbon nanotubes (CNTs) are materials showing properties favourable for field emission. CNT field emitters were fabricated to show potential for advanced use in space applications. This was done using using the electrophoretic deposition method and a current density of 10 mA/cm2 was obtained at an applied electric field of 6.14 V/μm.

Bernard Tashie-Lewis

Aerospace Engineering

Electronic 'Dragonball' Hide-n-Seek Game

Creating a magical game in a physical electronic world

Inspired by a Japanese Manga and Anime series of the same name, these artefacts follow closely to that fictional story. When the orbs are separate they will do nothing, but when they get closer together they will light up and pulse, and greater proximity increases brightness. When they are all collected together they light up and pulse in sync. The idea being that, if the orbs are hidden around a room, with one of them in the user's possession, they can find the others with the light as an indicator, an object form of hide and seek. The artefacts will communicate with each other using high frequency sound, and the embedded system within will measure the strength of the sound signal and respond with the light emitted from the LEDs.

Turgay Hassan
Product Design Engineering

Online and multimedia based concepts.

humanistic innovation   19

sustainable innovation   197

technical innovation   245

digital innovation   321

network directory   353

My Cupcake Cafe

Interactive database-driven recipe website

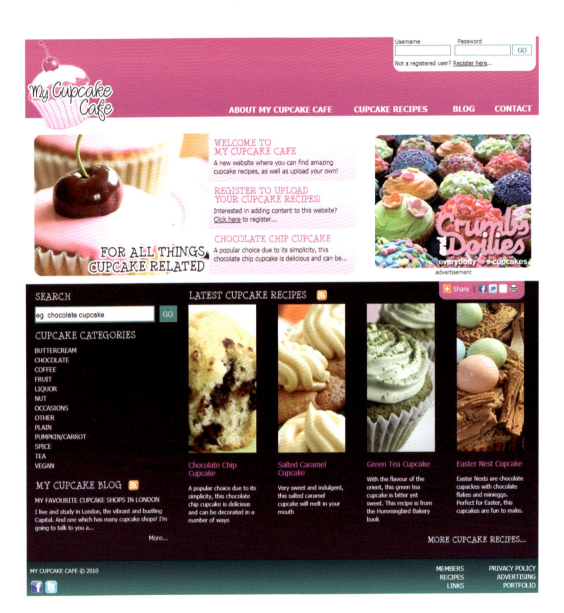

My Cupcake Cafe is an interactive database-driven recipe website which showcases cupcake recipes and cupcake decorating tips. The website counts on user added content to function through a built-in registration and content management system where an account can be created. From here, users can create a profile, add and delete their own cupcake recipes in categories, as well as adding their own cupcake related images and videos. The recipe pages are content rich with social networking and sharing tools, as well as the ability to comment and star rate a recipe. The CMS enables admin control over the website, ensuring moderation and verification. A blog is also present on the website to share recipes and cupcake baking and decorating experiences.

Daniela Albano

Multimedia Technology and Design

Simple Symphony is a website created to educate users about the fundamentals of instrument notation. The website is aimed at hobbyists, amateurs and higher education musicians to aid instrument selection, their tonal qualities and performance markings. It provides an interactive user experience, with the aim of encouraging the user to compose in a more sympathetic manner. The site hosts over 30 HD videos, which professionally deliver demonstrations of instrument articulations and composition techniques. Simple Symphony may be used as an additional teaching resource to accompany classroom learning. Interactive elements have been designed to be intuitive, visually engaging and user friendly to develop a deeper appreciation of the musical field.

Fiona Cullen
Multimedia Technology and Design

DrivingL

The online driving school: learning to drive the visual way

DrivingL is an online driving school website. The motivating concept behind the project is the use of interactive media to teach or enhance driving knowledge. This concept was fulfilled through detailed video tutorials aimed at teaching individuals the basics of driving a car. There are also interactive tests to accompany the tutorials for individuals who wish to assess their improvement. The project is also displays and showcases new skills and techniques that have been acquired through the process of this project. Elements such as branding, website design and video editing were used as well as features like a green screen presenter to enhance the interactivity of the website.

Anthony Ramnath

Multimedia Technology and Design

Future Driving Environment

Augmented reality interface with multi-touch gestures

V-Strike

Immobilization of a rogue vehicle for the emergency services

As technology progresses, our cars are exposed to an increasing amount of data. Connected vehicles, networks of cars, driving assistance, internet access and so on. These information touchpoints have the potential to significantly enhance our driving experience, but also pose a potential risk in complicating the cognitively demanding task of driving. This project aims to offer a solution to this contradictory future evolution through a non-intrusive heads-up display interface. The information is displayed to the driver on various tabs directly on the windscreen, depending on the environment surrounding the car. Driving data is ranked according to its priority and necessity: Rank one data is displayed instantly whereas Rank two data appears through a notification system. Third rank data is displayed only when the environment is secure (waiting at traffic lights for example). Thanks to an eye tracking system, the interface adapts itself to the driver's behaviour; parts are dimmed and magnified according to sight direction. The driver is able to manipulate these panels of information easily thanks to touch zones on the steering wheel; multi-touch gestures allow rich, intuitive and unobtrusive interaction. Such a system helps to deal with a large amount of data while preventing the confusing multiplication of controls.

The concept is based in the year 2020 where government legislation has been passed allowing the police force to take control of a member of the public's vehicle if they break the law and fail to stop during a police pursuit. V-Strike stands for 'vehicle strike' and is for use by the police force to gradually disable rogue or 'failure to stop' vehicles, through direct wireless communication (WLAN) to the rogue vehicle's engine control unit (ECU). V-strike sends a wireless signal/code which cuts all vehicle engine functions by 90%, allowing for a smooth, controlled and safe stop. V-Strike utilises contemporary display technologies for clear visuals, namely a targeting circle, which can be manoeuvred to target the rogue vehicle (like a missile lock in a fighter jet) and once 'locked on' the signal to the ECU can be sent (this will ensure the signal reaches its designated target). Each vehicle's ECU will be pre-built with a special algorithm which only police may access and will be pre-programmed into the device/product. Once activated it will override the manufacturers' ECU settings, giving police full access and ultimately control of the rogue vehicle.

Antoine Auriol

Integrated Product Design

Sebastian Fraser

Tendon Clinic UK

Patient information portal regarding tendon injuries and recovery

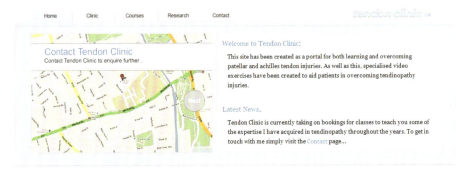

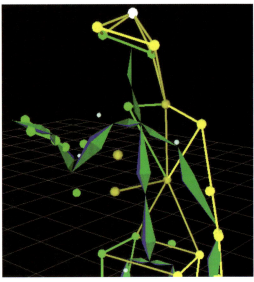

The company Tendon Clinic sponsored the creation of a website to depict information regarding the area of tendon injuries (tendinopathy) and information for their customers on how to overcome these injuries. Tendon Clinic also set the task of creating a company logo as well as the company website. This will be used as a portal for its customers to view information regarding injuries to the patellar and achilles tendons. Ultimately, Tendon Clinic wanted the site to host 6 different animated exercise videos which could be used by patients to overcome injuries they may have sustained. To fulfil these objectives, a website was created through the combination of flash and html. These videos used motion capture and 3-D technology encapsulate the exercises that Tendon Clinic wanted to demonstrate.

Carl Durler

Multimedia Technology and Design

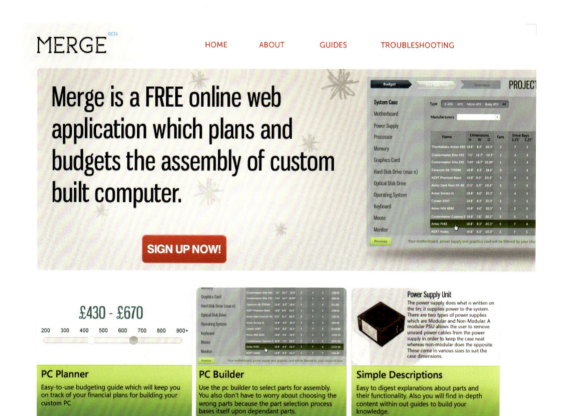

Merge is a web application which offers users a platform to plan and budget the assembly of custom PCs. The key feature of this site is the simple to use assembly wizard. The assembly wizard checks the compatibility of each component upon selection. Merge offers comprehensive tutorials on PC assembly for the experienced and inexperienced. A forum allows users to make queries about their assembly, installation, hardware, or other technical issues they may face. Lastly, users are provided with a unique profile page where they can view their saved projects, edit them and edit their details.

Richard Acquaye

Multimedia Technology and Design

Harrison Williams

Designing for Investment

Since the bubble burst in 2008, it has made the competitive market a real jungle. So, is it worth spending money on design?

With the toppling of powerful leaders, the playing fields have been levelled. Businesses are now searching for the most cost effective solutions to improve their products and services. In turn, this has reduced the prices of production and services to ensure that they retain their market share. So, is it worth investing your hard earned money in design, in the hope that it allows you to stand out from the crowd? Would this make you a market leader and capture customers from your competitors?

An obvious example of a successful company that has done nothing but succeed since the late nineties (with a share price rise of over 500%) and managed to dodge any major economic downfall in 2008, is Apple. Is it that Apple products are that much more superior to their competitors?

Frankly, no... if you look at the level of technology that Apple offers, for the prices that they demand, their competitors can offer similar, or in some cases, better technical specs for more or less the same money.

A Sony Vaio is a prime example of this. One thousand pounds will get you the base model MacBook Pro with a commendable technical specification, however when you spend the same money on a Sony Vaio, it will surpass the specification of the MacBook. So why do people pay more for less? It has been widely stated that this is down to the OSX software that Apple have developed which ensures that the Mac will outperform the competition. However, it has very little to do with either the specification or the software...

Back in 1997, Apple was in difficulties and Steve Jobs was reinstated as CEO. He shared a vision of what Apple should be in the future with a little known young English designer, Jonathan "Jony" Ive. Jony Ive was made the Head of Industrial Design and together they refocused the business on Human Centred Design. As a consequence, their products are based on the people who use them, not solely the hardware or software behind the product. The most obvious statement of this intent came with the production of the first iMac. In a time when computers were a cold, grey, box, Apple produced a colourful transparent desktop. They also integrated "fun" humanistic icons in the software; this also aided the softening of the boundaries between people and technology.

Apple has continued to follow this mantra, making simplistic products for "people", not just tech-heads. This opened up their market share, from a focused group of business professionals, to everyone. They have done this by investing hundreds of millions of pounds into the design of the whole experience. The design process has been so successful that Apple products are now fashion accessories, the "must have gadgets". Apple has continued to break the boundaries between technology and the people who use it. The latest obvious breakthrough has come with the incorporation of the touch screen on many of their products. Apple reinvented this technology by implementing the "pinch". The pinching motion to zoom in and out on products like the iPhone, iPad and iPod. This allows a tactile interface between the user and screen. This physical interaction means that you feel a part of the product and the functionality. It never occurs to you that there is a massive amount of time, money and effort invested in coding this action - it is just that "cool" thing that you like and are able to do!

This may be why the competitor phones are not selling as successfully. They have this function but also a completely different feeling... The feeling being that it took a lot of effort, money and time to code and you should be grateful that it works!

Apple offers their product as a Michelin star restaurant, hiding all the effort and strain, the long hours and the difficulties involved. Everybody wants to make a functional product, all singing, all dancing, to topple the mighty iPhone. As yet, the best contender is probably BlackBerry. Their direct approach to the phone as a business tool, based on function; completely reliable and well designed is proven. This is the time old argument against high design, what is more important: form over function, or function over form?

According to recently published figures, the Apple iPhone has sold 14 million handsets in the last quarter (Oct 2010) compared to 12.4 million handsets from BlackBerry. A difference of 11.4% is significant, but it also demonstrates that the iPhone is not a runaway winner; the public obviously has a need for a very functional product. This does in fact put the pressure back on Apple, as companies that are already experts in functionality are becoming more design orientated.However, it is clear to see what Apple have achieved by focusing their business around design. Steve Jobs said recently: "iPad is off to a terrific start, more people are buying Macs than ever before, and we have some amazing new products still to come this year."

You may not have a product-based business, so you may think there is no point in attempting to invest in design, that it is a waste of time and money. If so, you may be interested to find out that a research paper by Bruce Tether, ESRC Centre for Research on Innovation and Competition (CRIC) at the University of Manchester, noted that incorporating design, as the innovation for the business, increased the sales by 23% over five years.It becomes a simple issue of margins. If you class design as an advertising service, it is a cost to your business and therefore it could be classed as tax deductible.If you employ a decent branding agency this should only involve half a dozen meetings. Then it is down to you to implement the design and stick rigidly to the guidelines that the agency establishes.

This is the main reason that investing in design has worked so well for Apple. It comes down to one thing and that is brand consistency. Everything is slick and refined, from the product range to the advertising across all mediums. With the products they do it simply through the smooth and simple designs, accompanied by the intelligent use of material, which gives an aesthetic quality, finish and a basis for the structure. It is the same across their software interface. It is clean, crisp, and easy to use. The stores, apps and website are all the same. The comparisons are easy to make and have constituted the development of such a strong brand.

Of course, your business is unlikely to be as global as Apple. You may not have a multitude of stores worldwide, nor a multitude of platforms that support your business. But the principles remain the same. If you invest time and money in design, ensure that it is rigidly implemented and consolidate your company under one brand, it can make all the difference. It is the little things that can actually make the biggest impact, such as using the same fonts, in both your emails and PowerPoint presentations. These are the sorts of details that a good designer will ensure.

So cover your bases, invest in design and make sure that your competition stays two steps behind you!

Interactive Gesture Based Advertising

Interactive television advertising prototype with online analytics platform

A project to create an interactive television advertising platform that can track views, interactions and revenue in an online analytics platform. The platform is designed to be passive, with the intention of working with hand gesture recognition. When an interactive advert is played, viewers can choose to have information about the product/service sent to their inbox or have it added directly to their online account. An online analytics platform tracks viewing frequency, the number of interactions, and the generated revenue from the campaign. Using this data, companies are able to track and measure the success of campaigns, and plan future marketing activities.

Rishee Patel

Multimedia Technology and Design

Paste Up is a web content management system that gives the owners of small websites the tools they need to have full control of their online content. By employing an intuitive 'click to edit' interface, users can arrange text, images and media within the web page itself. Many CM systems abstract content into an overly complex and daunting interface that takes time to set up and learn. Paste Up can be rapidly deployed and, by borrowing semantic aspects from word processing software, is instantly accessible to anyone with basic computer literacy and design skill. The aim of the project is to close the gulf that exists between the front-end of a website and the editing of its content by removing this abstraction layer and making content management straightforward and intuitive.

James Shakespeare

Multimedia Technology and Design

uVisionEPG
An electronic programme guide for 3D-TV

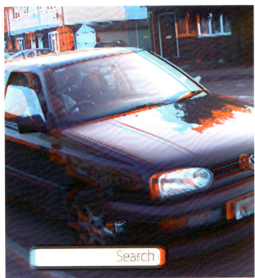

HDTV was so 2009! 3DTV had an impressive year in 2010 with growing interest from content developers in both TV and Film (Sky Football, Avatar 3D), and an increasingly large range of different 3D-TV systems to buy. 2011 boasts even more technology to choose from, lower prices and yet more content. The only thing missing is the 3D Interface; specifically the Electronic Programme Guide, EPG, which is where uVisionEPG comes in. This concept aims to retain tried and trusted features of previous EPG systems, transforming them using 3D space in order to enable all age groups to enjoy this new technology and improve the overall user experience. uVisionEPG allows keyword searching, personalised channel lists, and the ability to scroll through time in a 3D environment making for a more enjoyable and immersive experience.

Trevor Swayne
Multimedia Technology and Design

From historical collections large and small, Culturebook brings together the planet's heritage, documentation, interpretations and perspectives. This wealth of content is delivered in a single smartphone application for education and contribution. It is an interactive visual/audio tour guide for all museums in your pocket. The application consists of a functional user interface that demonstrates the overall concept. The main areas of demonstration are exhibit profiles, including links to multimedia such as related podcasts, videos and user content submissions. Content could potentially be personalised to different users based on interests, age and so on. The exhibits could be collated by the user to create a new synthesis and be shared via social media.

Calum Ryan
Multimedia Technology and Design

Roll Tomorrow

A film combining 8mm film stock with HD digital footage

The concept of this project was to combine 8mm (Standard 8) film footage with HD digital footage in order to create a film trailer. The film is set partly in 1950's America and partly in the modern day, with the 8mm footage used to represent the former and the HD footage used to represent the latter. The project revolves around satirising the genre of teenage rebel films from this era and presents this through a faux trailer for a nonexistent, conceptual film. The story follows the exploits of Johnny LaCleve, a 1950's 'juvenile delinquent' who is stereotypical of a character often found in these movies.

Keahn Rahimi
Broadcast Media

'Below' is a short 3D animation about a man who, after believing what he reads in the headlines about swine flu, descends underground in an attempt to escape what he perceives as his impending doom. It tackles the issue of newspaper sensationalism which has had relatively little coverage in popular media and is becoming more and more prevalent in our society. The film is without dialogue so relies on camera animation and the atmosphere created by the environments to convey the theme.

Laurence Dawes
Multimedia Technology and Design

Washers & Sundries
A short film

Washers & Sundries is a short film using a mixture of both animation and live video. The film itself focuses on three hand drawn sketches, created by a young girl, which magically come to life off the page. The narrative focuses on the paper characters' interactions with both their environment and human counterparts before the characters leave to explore the wider word. The film uses a variety of technical aspects such as animation, compositing, motion tracking and audio synchronisation. The video content was shot using a DSLR (Canon 550D) and features characters created and animated digitally, rather then hand drawn animation.

Mike Malone
Broadcast Media

This animation uses a combination of 2D and 3D techniques to focus on the use of emotion in media, in order to understand how emotions are communicated and used to increase the engagement of the audience. The design of the animation had to be aesthetically pleasing whilst demonstrating a particular illustration style and incorporating the overall theme of the project. The project also including producing the sound to accompany the animation, recording sound effects and sourcing music where necessary.

Chi Yu

Multimedia Technology and Design

Dreams
Experiences and interactivity

'Dreams' explores the relationship between dreams and waking experiences in the form of an animation utilising various visual styles as well as the idea of using interactivity, so that users have some control and take away a better experience. The animation contains various movie clips taking place during three days in the life of a character (Josh) and also showing the dreams he has when he sleeps. Josh has problems that he must deal with, since he has an exam on the third day and unless he solves them he may not wake up in time, since the problems have been causing him to lose sleep. Interactivity is introduced at the end of each movie clip, where the user has to make a decision which affects the later day and dream scenes, reflecting decisions we experience in real life.

Seán Grannum
Multimedia Technology and Design

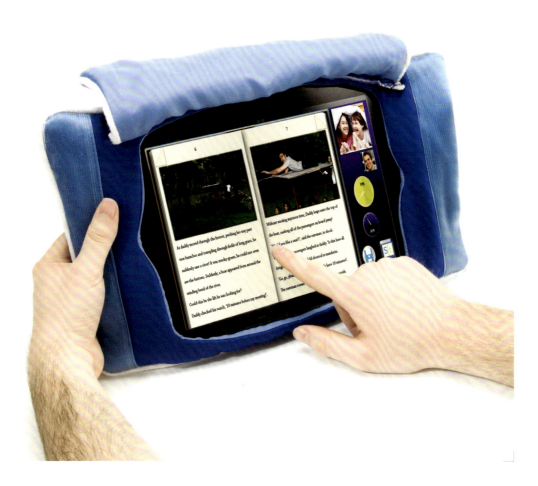

As our communities grow wider and spread throughout the world we rely upon impersonal and anonymous digital communication as a means of sharing emotions with those we miss. This concept provides a communication medium that allows for natural expression between loved ones through the act of storytelling. Storytelling is a ritual missed by many a parent who spends time away travelling for work. Each party will use software integrated within a wireless tablet device to capture inspiration around them. The software will generate immersive stories to be shared at bedtime. This ritual is heightened through a plush accessory providing interactive elements. It also doubles as a travel pillow, making it efficient to travel with. Tales from their journey will inspire their children whilst helping them feel closer to the ones they love.

Steven Gutteridge
Industrial Design and Technology

Roman Luyken

The Holistic Conceptualist

For the last century we have been living in an increasingly prescriptive and technologically driven product culture, and as a result we live in an abundance of efficiently working technology capable of solving most problems. In the past this has worked well for us, but with computers replacing the traditional knowledge worker and an economic crisis, it is time for change. The profile of a holistic conceptualist is likely to increasingly shape the future professional.

Emotional Intelligence

In a career-driven culture we have lost touch with what is important to us as humans, especially at an emotional level. The desire to earn a good living may have driven us to pursue professions such as medicine, law and engineering, which required sequential, linear and logical thinking. Today, a computer easily outperforms the human brain in speed, accuracy and cost, pushing us to change the way we work and live, but also allowing us to reconsider what we want in our lives, both as creators and consumers. It is at this point that we are realising the value of our brain's logical counterpart and former 'threat', the emotional centre. Emotional intelligence is what makes us human, what differentiates us from computers. Emotional intelligence is 'simultaneous, metaphorical, aesthetic, contextual and synthetic', it lets us think laterally, non-linear, creatively and, of course, emotionally. It is this kind of thought which allows us to interact with, and to understand people as well as to see the big picture, to absorb seemingly unrelated events, to understand randomness and bring all these elements together. The holistic conceptualist employs both: logical, knowledge-based, intelligence as well as emotional intelligence in utilising the large pool of information and technology available to us by re-contextualising it in an original way that connects with people.

Re-contextualising Emotion into the Modern Business World

"A deep understanding of people is the key to success in today's market place" Pat Jordan.

What are people about and how does this relate to the way a conceptualist thinks? Dan Pink categorizes our emotional side into six 'high-touch' senses defining the conceptualist: design, story, sympathy, empathy, play and meaning. Pat Jordan describes what drives us humans into physiological, sociological, psychological and ideological pleasures and Don Norman breaks up our levels of thinking into visceral, behavioural and reflective. Although phrased and categorised differently, in essence all these approaches point in one direction: we as people are all about emotions. What we feel and how we feel, what drives us, what determines the way we react and ultimately what differentiates our lives from being fulfilled or not. The emotional centre of our brain is in charge of this, and it can be influenced by the tiniest nuances. Being a holistic conceptualist is about understanding emotional behavior in its nuances and catering to its needs.

With technological requirements fulfilled, there is now space and time to indulge emotional needs. A prime example for this is the car industry, in which all cars run efficiently, and the main differences between brands lies in the user experience. Take a look at the BMW Z4. This car is not about being an automobile, but rather about the induced aesthetic appreciation, the feel of the leather against your skin, the engine sound, the detail in design, all brought together in symphony to create an "expression of joy" to be experienced by the driver. This principle has been demonstrated in smaller consumer line cars such as the Fiat 500 and the new Mini and the success of such attributes is now self evident.

Cost is always a factor, and how much the majority of people are willing to pay for such a "superfluous luxury" is debatable. Luckily, we live in a world of technological and infrastructural advancement in which conceptualistic emotional design becomes accessible at virtually no cost, as demonstrated by the Facebook business model. The whole new approach to creating highly profitable businesses by offering products and services for free, was definitely not created by the minds of purely logical thinkers but rather those of conceptualists who have given their emotional brain-halves trust and freedom. It is interesting how 'modern' companies such as Google have introduced new working schemes such as 'FedEx' days, in which employees are sent out to do as they please to then return with one original idea of any kind. Surprisingly enough, some of the most profitable ideas have originated from these or similar schemes in which pressure is taken away from employees and creative thinking is encouraged.

The Lucky iPod Generation

For the iPod generation, born in the late 20th century, the iPod in itself, created by conceptual thinkers to emotionally bind with and stimulate people through design and music, was quintessential in also forming the future conceptualist. Rather than being a distraction to one's focus, it has encouraged the iPod generation to utilize and understand both sides of our brain. The iPod generation conceptualist, a digital native, has grown up using technology to boost his or her logical processing power seamlessly and effortlessly, leaving space and time to develop emotionally centered creative ideas.

The economic crisis comes as a blessing to the iPod generation conceptualists who are keen to start new ideas from scratch without being tied

"A deep understanding of people is the key to success in today's market place" Pat Jordan

to the laws of their forefathers. The combination of a recession in which there are no 'wrong' approaches and the clean and undamaged mind of a graduate who has nothing to lose, are an inspirational source of high potential. The world is open to a new era to be formed in new ways.

Emotional Intelligence can handle more variables than logic, allowing one to reach conclusions when faced with complex problems. We have all experienced trying to solve a difficult decision in which every thought creates more and more complex variables, making it virtually impossible to decide on pure reason. At this point emotional intelligence can be called in to decide. Although a 'gut' decision may seem irrational, the emotional decider still takes all information ever fed to the brain into account, gauging personal preferences and impressions against pure reason and logic to come up with a result superior to a purely 'logical' decision, as these often do not take elementary 'Black Swan' events into account. The 'Black Swan Event' identified by Nassim Taleb is a seemingly insignificant factor which can turn out to be responsible for detrimental events of disproportionally high significance. Emotional lateral thinking will have a higher chance of taking such an event into account while logic would discard it immediately.

The Future of Online Dating
Relationships in a connected society

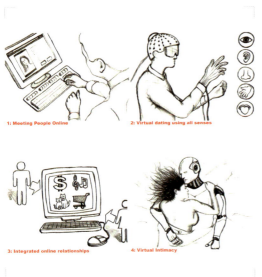

1: Meeting People Online   2: Virtual dating using all senses

3: Integrated online relationships   4: Virtual Intimacy

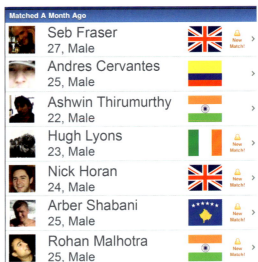

'The Future of Online Dating' is a project that evolved from the stereotypes of dating, relationships and intimacy. Future is now being directed towards virtualization of societies, and computers are being placed at the heart of most operations. It is not about making arbitrary guesses, but systematically analysing past and present data, both subjectively and objectively, in order to minimise uncertainty and develop an appropriate strategic direction for both the near and distant future.

We conducted an experiment whereby we constructed a fictional dating profile, and watched it far outperform an 'honest' profile written by the same subject. It was interesting to observe how the lack of transparency in the industry manifests itself, and whilst it may be beneficial in the short run to 'daters', in the long run it is likely to cast a shadow over the industry, leading to a degradation of image and rendering brands worthless.

The classification of relationships into incidental relationship, purpose-driven relationship, friendship relationship and lifelong relationship offers insights into the emotional aspect of product segregation by its purpose, which was very useful while designing for virtual intimacy. Wally Olins did however raise another interesting point - that, in contrast to his overriding view, it is possible that it is the individual's personal brand which they destroy through dishonesty, rather than the brand of the site they represent. Human behaviour and relationships are extremely unpredictable and complex interactions and as relationships move into the virtual world, in the future design will have to play a crucial role in ensuring that the intimacy and honesty of physical contact can be safely replicated in the virtual world.

The scales for filtering topics for project definition was presented by Kevin McCullagh, in his illustration of Tim Brown's 'Design Doing' which classifies design to be a combination of Desirability, Viability and Feasibility, along with the suggestive themes of clarity, style, authority, co-ordination and idea blue sky thinking proved efficient and served as a catalyst of innovation.

In addition to using conventional methods like academic and online research, getting expert opinions, brainstorming, creating future scenarios, sketching etc, this was an excellent opportunity to sharpen our design thinking skills.

As blue sky concepts the team came up with the idea of being able to take finding dates online even further, including having relationships and being intimate via the web. The foundation of this idea comes from the fact that many people believe their future partner could be anywhere in the world, and have limited opportunities to find a potential partner.

A. Cervantes, S. Fraser, N. Horan, H. Lyons, R. Malhorta, A. Shabani, A. Thirumurthy
Design and Branding Strategy

Happy Clean

Clean. Earn Points. Get Reward

we:care

A 'Big Society' community based
social care service

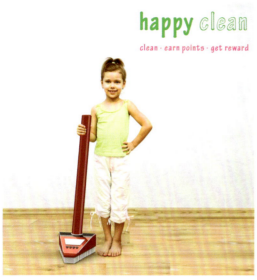

Imagine role play meets Nando's grilled chicken reward scheme meets reality, using this analogy 'happy clean' is a project create to encourage children to help with household chores. Personality development continues from birth to maturity and the aim was to assist this through the use of a product that improves parent–child interaction. Children are often reluctant to contribute to parents' workload due to the associated perception of boredom and tedium. 'Happy clean' bridges this gap by introducing interactive features to a conventional mop to add interest for young people. An in-built timer records the time the child spends with the product and the points earned. Additionally, the timer can be used to set a target time within which the child should complete an activity in order to earn extra points. All data is stored within 'happy clean' and once connected to a computer, the kids, in accordance to the points scored, may choose from rewards offered by parents, such as toys, clothes, treats and so on. The child develops a sense of gratitude towards the parent, positive associations with housework, and the parent can appreciate the responsibility developed by the child in their formative years.

Increasing funding gaps, poor user experiences of social care and an ageing UK population all demand social innovation in the public sector. 'we:care' is a 'Big Society' community-based care service that encourages local people to be involved in social care across their community. The service simplifies ways in which community members can help each other whilst giving incentives to do so. In concept, people sign up for a 'we:care' account, connecting them to their local 'we:care' community through multiple touchpoints (including a smartphone app, website and membership kit complete with smartcard). Requests for help from 'we:care' members can then be entered, searched and accessed. Responding and providing help earns 'we:care' members' points, which can be spent at local shops, on travel concessions, or simply saved for the future when help is required. The 'we:care' service design offers a range of synchronised touchpoints as options for inclusive user populations. 'we:care' aims to create a more financially sustainable community-based care system to reduce wasted funding and resources and to encourage caring within local communities, ultimately improving community cohesion. The project recently won a funding contract with the Design Council/TSB in answer to the 'Independence Matters' brief and it is currently being developed in collaboration with social innovation consultancy FutureGov.

Ashwin Thirumurthy

Integrated Product Design

Murtaza Abidi

More – The Little Time Saver
A device to help manage time consumption for the user

CIRS
Community interaction system

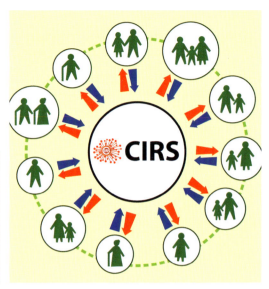

In today's world countless numbers of technological gadgets surround us, providing us with luxuries we could not have imagined 10 years ago. Smartphones, netbooks and tablets mean that we can carry entertainment and web access with us everywhere. However, the problem remains that there is a lack of help with the simplest things we do in our daily life. "More" is a wearable technology that connects to all a user's devices, sources of information such as reminders, meetings, to-do lists and displays the desired information at the right place and time. With built-in location-aware GPS technology, it displays grocery lists as the user enters a store, or work to-do lists only during work hours. Sleep-monitoring technology means that "more" knows when the user wakes up and can present useful information according to the time of day. At the start of the day information such as weather forecasts, daily to-do lists or even bills due that day are displayed. The wrist display interface aims to free the user from the stress of trying to remember every detail of their hectic daily schedules. "more" deals with the small things in life that tend to take up the most time.

There are several consequences of a compact living style such as social and health problems. To combat such issues, a 'Community Interaction Reinforcement System' (CIRS) was designed: a sharing system for the residents of a compact lifestyle community e.g. council flat and condominium. CIRS acts like a virtual front yard for residents to connect with their neighbours, to share ideas, information, or equipment. CIRS also allows the residents to ask for help, advice, or borrow things from their neighbours. It is a channel to set up the activities within the community. Residents can access CIRS via three channels: mobile application, internet webpage, and interactive access point located at the main entrance of the building. The access point consists of a touch screen and an interactive LCD board inclusively designed for all residents who cannot access the system via other channels. The interactive board displays news, emergency alerts, information posted by the residents, and general information like traffic reports and weather forecasts. Each time the residents access CIRS, credits will be collected in their accounts. These credits can be redeemed for tax reduction or discount coupons. This system will boost interaction between residents, improve quality of life, and increase security in a compact community. It helps create a self-sufficient community, which can reduce social problems.

Sarp Suerdas

Nopadon Chaleampao
Integrated Product Design

Virtual Reality Storytelling

Social interface reducing intergenerational differences

Enjoy Everyday

Communication aid for older people

Our ageing population is seeing a parallel increase in social exclusion due in part to negative stereotypes about the elderly. This concept aims to prevent the development of negative stereotypes, attitudes and perceptions in the early stages of childhood using the grandparent-grandchild relationship. Virtual Reality (VR) is used as a platform for improved interaction and reduction of intergenerational differences. The product builds on traditional storytelling using a book as the control leading to an intuitive user interface, while adding an immersive VR experience to increase the quality of the experience and counteract disparities such as mobility, cognitive ability and overall health. The grandparent leads the grandchild through a number of experiences where knowledge transfer, bonding, trust and understanding are the basis of the activities. Using avatars the grandparent and grandchild interact in the virtual space. The first stage allows the grandparent and grandchild to swap avatars, experiencing life in each other's shoes and developing a different outlook, showing each the experience of another generation and reducing the development of negative stereotypes. Lastly, it transforms the interaction by filtering or augmenting attributes that negatively affect the interaction, while allowing for the provision of extra information or cues in the grandparent's display further improving the exchange of knowledge, entertainment and bonding.

Research has shown a number of strong links between ageing and depression. The concept of Enjoy Everyday aims to address helplessness as key symptom of depression in older people. The system offers a new platform for users to share their stories and communicate these insights to other people in society. The design uses two separate interfaces for input and output. The design of both interfaces aim to make interaction easy for older users by focusing on principles of inclusive and interface design. It uses voice recognition technology so that users need simply talk about their experiences. Simultaneously images are attributed to stories as they are recorded in order to stimulate memory. The output interface acts a medium for the general public to share the recorded stories. An integrated camera recognises faces and selects people to share particular stories with.

Hugh Lyons

Integrated Product Design

Haoyun Xue

Mixitup UK

Highlighting segregation within London

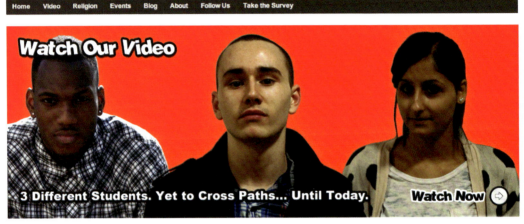

Mixitup UK is a website for young adults, to introduce to them to the diverse cultures of London. The aim of Mixitup UK is to bridge the segregation gap, and encourage young adults to learn and engage with different cultures. The idea came from witnessing the different cultures within the university and the levels of interaction, or lack thereof, present between them. People tend to form social groups with people who are similar to them, be it by race or religion. Mixitup raises awareness about the importance of integration through a short interactive drama. It teaches young adults about London's diversity, in order for them to be able to celebrate their differences. It encourages young adults to attend different cultural events, in order to engage with different people.

Ansbert Dodoo

Multimedia Technology and Design

ANNOTATE THAT!

Annotate That! is a brand new way to share and express your opinions via the web medium. A user of Annotate That! can leave annotations on a web page, image or document. He or she can send and/or share these annotations from a unique link with anyone in the world. Annotations build up progressively, even into discussions. If you have access to the link then you can leave further annotations or comments. Please visit www.annotatethat.com to use the application.

Dean Claydon
Multimedia Technology and Design

Eco-Warrior
HTML5 web app

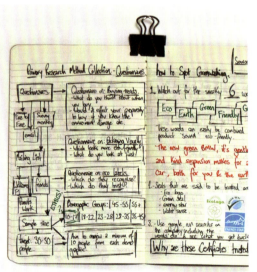

Do modern day consumers realise the environmental impact of their purchases? Eco-Warrior has been created in response to the issue of the consumers' lack of environmental clarity. By educating the consumer about the environmental consequences of their purchases, it is hoped they will make more informed and responsible purchasing decisions. With the smartphone now an inherent part of the buying process, it presents an excellent medium to provide impartial environmental product information, previously not commercially feasible. Eco-Warrior works by providing perspective to unquantifiable technical product specs, showing eco-labels to trust and brands to believe in. It offers an impartial look at a product's environmental claims and encourages the consumer to make more responsible choices.

Michael Stanley
Industrial Design and Technology

Scrapped

An online marketplace for reduction of electronic waste

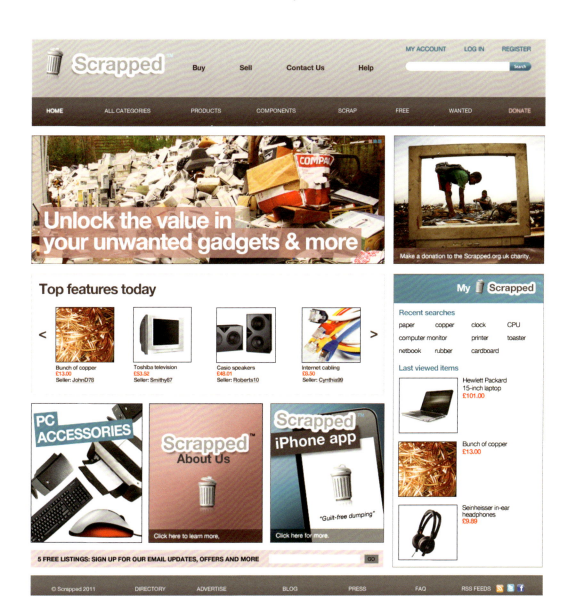

Computers, mobile phones, televisions – everyone owns them, but the average person does not think about their life beyond the dump. Electronic waste is one of the fastest types of waste. Undeclared, old electronic goods are routinely exported from the West to developing countries. Hazardous e-waste violates international laws and poses a huge health risk to people in the affected communities. Scrapped.org.uk is a charity and online marketplace where people can buy and sell used electronic goods and more. Aiming to raise awareness on the e-waste issue, the website ultimately makes you think twice about the purchase and disposal of gadgets. This is only just the beginning toward a solution for reducing e-waste.

Shirley Denchie

Multimedia Technology and Design

The people.

humanistic innovation   19

sustainable innovation   197

technical innovation   245

digital innovation   321

network directory   353

1. **Mohamed Abdel-Gadir**
Product Design Engineering BSc
moodeedesign@gmail.com
moodee.mdotmedia.com
07909 441873
Page: 140

2. **Murtaza Abidi**
Integrated Product Design MSc
murtaza.s.abidi@gmail.com
www.murtazaabidi.com
07863 555 051
Page: 343

3. **Richard Acquaye**
Multimedia Technology & Design BSc
richard.acquaye@hotmail.co.uk
richarda.deviantart.com
07944 459287
Page: 327

4. **Nadim Al-Twal**
Design & Branding Strategy MA
altwal.nadim@gmail.com
www.nadimtwal.com
07592 828222
Page: 191

5. **Daniela Albano**
Multimedia Technology & Design BSc
info@neondaniela.co.uk
www.neondaniela.co.uk
07846 396 919
Page: 322

6. **Lucy Alder**
Product Design BSc
luc.alder@gmail.com
www.lodilea.co.uk
07764 679243
Page: 112

7. **Ashima Amar**
Design & Branding Strategy MA
ashima.amar@gmail.com
07740 973469
Page: 172

8. **Alexander Ambridge**
Product Design BSc
design@alexambridge.co.uk
www.alexambridge.co.uk
07816 822442
Pages: 206, 259

9. **Nazanin Amjadi**
Integrated Product Design MSc
Page: 194

10. **Dave Anderson**
Product Design Engineering BSc
info@creativeengineer.co.uk
www.creativeengineer.co.uk
07872 002326
Page: 258

11. **Steven Arding**
Aerospace Engineering MEng
steve_arding@hotmail.co.uk
07917 834360
Page: 312

12. **Kylie Arthur**
Design & Branding Strategy MA
kylie_arthur@hotmail.com
07956 922 202
Pages: 176, 191, 193, 194

13. **Kourosh Atefipour**
Product Design Engineering BSc
kouroshatefipour@yahoo.co.uk
07837 652992
Pages: 265, 279

14. **Antoine Auriol**
Integrated Product Design MSc
antoine.auriol@me.com
www.antoine-auriol.com
07869 513518
Page: 325

15. **Seun Babatola**
Industrial Design & Technology BA
stola_148@hotmail.com
07980 202250
Pages: 98, 285

16. **Penny Bamford**
Integrated Product Design MSc
pennybamford@yahoo.com
07590 444602
Page: 231

1. **Sorana Barbalata**
Industrial Design & Technology BA
office@sorana.co.uk
www.sorana.co.uk
07877 969485
Pages: 58, 213

2. **Julian Bennet**
Aerospace Engineering MSC
Page: 228

3. **Stephen Blakeney**
Mechanical Engineering with
Aeronautics BEng
stephen@blakeneys.co.uk
07525 465271
Page: 229

4. **Jade Boggia**
Product Design Engineering BSc
jadeboggia@hotmail.co.uk
www.jadeboggia.co.uk
07810 435148
Pages: 110, 282

5. **Yinka Branco-Rhodes**
Industrial Design & Technology BA
infront@live.co.uk
infrontyinks.zapto.org
07702 340400
Pages: 92, 142

6. **Edmund Bright**
Mechanical Engineering MEng
edbright1000@yahoo.co.uk
07796 315660
Page: 297

7. **Simon Brockie**
Civil Engineering with
Sustainability MEng
simon.brockie@gmail.com
07964 424320
Page: 214

8. **Ashley Brooks**
Design & Branding Strategy MA
abrooks.designs@gmail.com
www.coroflot.com/ashbash
07769 223558
Pages: 180, 192, 194

9. **Olly Brown**
Product Design BSc
oliverhbrown@me.com
www.oliverhbrown.com
07921 829815
Pages: 51, 123

10. **Brunel Masters
Motorsport**
bmm07@gmail.com
Page: 302

11. **Brunel Racing BR12**
Page: 300

12. **Elena Bruno**
Product Design BSc
elena.a.bruno@gmail.com
www.twitter.com/#!/b_Design_tweets
Page: 212

13. **Heather Bybee**
Design & Branding Strategy MA
bybeehe@gmail.com
www.coroflot.com/bybeehe
07580 245768
Pages: 184, 188, 193, 194

14. **Madeleine Carver**
Product Design BSc
madeleinecarver@gmail.com
07707 075659
Pages: 158, 224, 253

15. **Fatos Ceren Tan**
Design & Branding Strategy MA
crn.tan@gmail.com
07749 413373
Page: 182

16. **Andres Cervantes**
Integrated Product Design MSc
afcluna@hotmail.com
07581 196137
Pages: 181, 342

1. **Nopadon Chaleampao**
Integrated Product Design MSc
nopadonch@hotmail.com
07703 565166
Page: 344

2. **Jason Cham**
Industrial Design & Technology BA
jasoncham@hotmail.co.uk
07872 172071
Pages: 264, 288

3. **Nathan Chan**
Industrial Design & Technology BA
wingjay_ccw@hotmail.com
07979 626567
Pages: 56, 298

4. **Yao-Wen Chang**
Design Strategy & Innovation MA
Page: 241

5. **Yu-Hsin Chien**
Design Strategy & Innovation MA
Page: 241

6. **Grace Minjoo Chung**
Design & Branding Strategy MA
Page: 194

7. **Michael Chung**
Industrial Design & Technology BA
chungyuktao@hotmail.com
07954 541956
Pages: 256, 266

8. **James Clarke**
Industrial Design & Technology BA
jamesclarke1988@gmail.com
www.jamesclarkedesign.co.uk
07721 028749
Pages: 236, 292

9. **Timothy Clarke**
Mechanical Engineering with Auto
Design MEng
timothy.i.m.clarke@gmail.com
07759 050751
Page: 297

10. **Dean Claydon**
Multimedia Technology & Design BSc
claydondean@googlemail.com
www.deanclaydon.co.uk
07913 525490
Page: 347

11. **David Cole**
Product Design BSc
d_designs@live.co.uk
07930 614782
Page: 138

12. **Tom Collett**
Industrial Design & Technology BA
tom@bigorangedesign.co.uk
www.bigorangedesign.co.uk
07734 355520
Pages: 148, 242

13. **Mark Connor**
Industrial Design & Technology BA
mark_connor_design@Live.com
07784 292500
Pages: 101, 254

14. **William Couch**
Mechanical Engineering MEng
williamcouch14@hotmail.com
07919 844758
Page: 307

15. **Fiona Cullen**
Multimedia Technology & Design BSc
fiona_cullen@me.com
www.mei-ling.co.uk
07984 965948
Page: 323

16. **Michael Daly**
Aerospace Engineering MEng
michaeldaly316@yahoo.co.uk
07833 35624
Page: 315

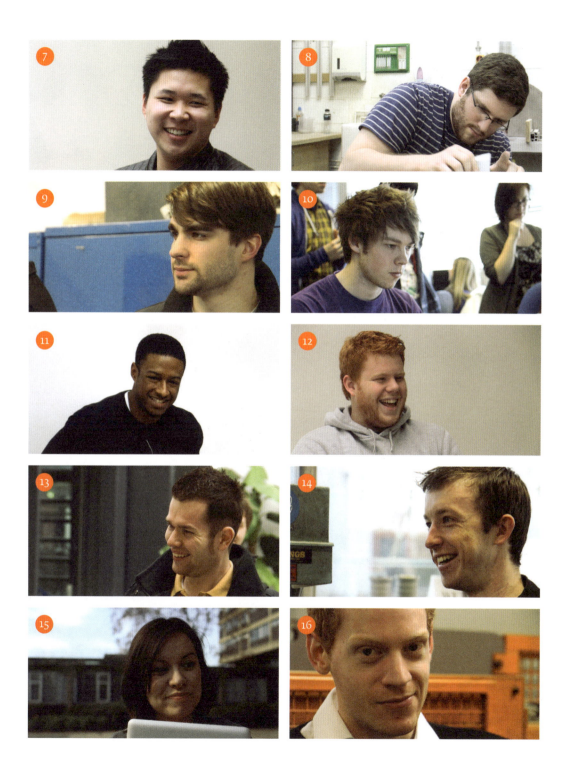

1. **Laurence Dawes**
Multimedia Technology & Design BSc
laurence.dawes@gmail.com
www.laurencedawes.co.uk
07891 077551
Page: 335

2. **Michael Day**
Product Design Engineering BSc
mikeypday@googlemail.com
www.empedesign.co.uk
07721 696939
Pages: 152, 286

3. **Shirley Denchie**
Multimedia Technology & Design BSc
hello@shirleydenchie.co.uk
www.shirleydenchie.co.uk
07947 936782
Page: 349

4. **Hannah Devoy**
Industrial Design & Technology BA
hannah.devoy@yahoo.co.uk
www.hannahdevoy.com
07825 077705
Pages: 40, 274

5. **Ross Dexter**
Industrial Design & Technology BA
rossdexter89@gmail.com
www.rossdexter.co.uk
07707 122735
Pages: 50, 237

6. **Jugbir Dhillon**
Aerospace Engineering MEng
jaggy_dhillon@hotmail.com
07965 645421
Page: 315

7. **Onkar Dhillon**
Aerospace Engineering BEng
dhillonxs@hotmail.com
07595 838792
Page: 316

8. **Amelia di Palma**
Industrial Design & Technology BA
ameliadipalma@hotmail.co.uk
07837 044722
Page: 139

9. **Juan Diaz del Castillo**
Mechanical Engineering with
Automotive Design BEng
juan.m.diazdelcastillo@gmail.com
07868 183758
Page: 304

10. **Ansbert Dodoo**
Multimedia Technology & Design BSc
ansbert_d@hotmail.co.uk
07958 008140
Page: 346

11. **Haoqiong Dong**
Design & Branding Strategy MA
donghaoqiong@hotmail.com
07403 531489
Page: 190

12. **Timothy Dunkley**
Product Design BSc
tradunkley@yahoo.co.uk
www.timdunkley.co.uk
07769 680617
Pages: 117, 162

13. **Carl Durler**
Multimedia Technology & Design BSc
carlito_is_here@hotmail.com
07535 491201
Page: 326

14. **Goldbert Dy**
Aerospace Engineering BEng
admiral_gdy@hotmail.com
020 87435549
Page: 316

15. **Robin East**
Aerospace Engineering MEng MEng
robineast@hotmail.com
07515 343861
Page: 315

16. **Nicholas Edgar**
Industrial Design & Technology BA
hello@edgar-design.co.uk
www.edgar-design.co.uk
07894 734849
Pages: 25, 165

1. **Tom Ellson**
Mechanical Engineering MEng
puma_arai@hotmail.com
07533 481 394
Page: 296

5. **Jon Fletcher**
Industrial Design & Technology BA
jon.s.fletcher@gmail.com
07507 160490
Pages: 44, 62

9. **Abraham Garcia**
Civil Engineering with
Sustainability BEng
ab_garci@hotmail.com
07504 896871
Page: 295

13. **Steven Gutteridge**
Industrial Design & Technology BA
stevengutteridge@gmail.com
www.spgutteridge.com
07554 174841
Pages: 63, 339

2. **Mohammed Elsouri**
Product Design BSc
elsourim1@hotmail.com
07999 191378
Pages: 26, 153

6. **Sebastian Fraser**
Integrated Product Design MSc
sebfraser@hotmail.co.uk
07792 371437
Pages: 325, 342

10. **Qing Ge**
Design & Branding Strategy MA
purple_air131@hotmail.com
Page: 190

14. **Hannah Haile**
Industrial Design & Technology BA
haile_88@hotmail.co.uk
www.h2-design.com
07854 538991
Pages: 100, 164

3. **Farzam Fahimi-Rad**
Aerospace Engineering BEng
Page: 316

7. **Luiza Frederico**
Design Strategy & Innovation MA
luizabfrederico@hotmail.com
07789 261385
Pages: 185, 188

11. **Iulia Gramon-Suba**
Design & Branding Strategy MA
iuliagramonsuba@ymail.com
07935 855122
Pages: 175, 192, 194

15. **Claire Hall**
Design & Branding Strategy MA
claire_hall@live.co.uk
07506 687664
Pages: 183, 191, 194

4. **Dee Fisher**
Product Design Engineering BSc
dekanifisher@yahoo.co.uk
07779 422114
Pages: 76, 94

8. **Grace Gana**
Design & Branding Strategy MA
gana.grace@gmail.com
07580 466293
Page: 191

12. **Seán Grannum**
Multimedia Technology & Design BSc
sean.grannum@gmail.com
alimitessmind.tumblr.com
07545 099391
Page: 338

16. **Muhammad Hamid**
Aerospace Engineering MEng
nomi.hamid@hotmail.co.uk
07852 942155
Page: 315

1. **Victoria Hamiliton**
   Mechanical Engineering MEng
   vrh0612@googlemail.com
   07917 015177
   Page: 33

2. **Richard Harris**
   Industrial Design & Technology BA
   richard.t.b.harris@hotmail.co.uk
   07828 922844
   Pages: 23, 275

3. **Turgay Hassan**
   Product Design Engineering BSc
   hello@turgayhassan.co.uk
   www.turgayhassan.co.uk
   07702 134945
   Pages: 116, 318

4. **Katie Hassell**
   Aerospace Engineering MSC
   Page: 228

5. **Katie Henbest**
   Industrial Design & Technology BA
   katiehenbest@gmail.com
   07545 785539
   Pages: 39, 122

6. **Kai Yeung Ho**
   Aerospace Engineering MEng
   samho1234@hotmail.com
   07810 295606
   Page: 315

7. **Adrian Hodder**
   Product Design BSc
   design@adrianhodder.co.uk
   www.adrianhodder.co.uk
   07913 161108
   Pages: 108, 257

8. **Emmanuel Hope**
   Product Design BSc
   ehope@live.co.uk
   www.ehopedesign.com
   07947 254583
   Page: 284

9. **Nicholas Horan**
   Integrated Product Design MSc
   nwhoran@hotmail.com
   www.horandesign.com
   07709 307322
   Pages: 52, 194, 342

10. **Szu-Chuan Hsu**
    Design Strategy & Innovation MA
    Page: 241

11. **Nicholas Jackson**
    Aerospace Engineering MEng
    jicholasnackson@gmail.com
    07830 238694
    Page: 312

12. **Tom Jay**
    Product Design BSc
    info@tomjaydesign.co.uk
    www.tomjaydesign.co.uk
    07738 098475
    Pages: 204, 255

13. **Victor Jeganathan**
    Product Design BSc
    vicjeg@googlemail.com
    07525 133243
    Pages: 272, 280

14. **Shaochen Jiang**
    Design Strategy & Innovation MA
    Page: 241

15. **Philip Jones**
    Mechanical Engineering BSc
    phil_jones_33@hotmail.com
    07595 583660
    Page: 226

16. **Tom Jones**
    Industrial Design & Technology BA
    ThomasJon3s@gmail.com
    07824 881090
    Page: 249

1. **Noppan Kaewkanjai**
Integrated Product Design MSc
tor_econ@hotmail.com
07958 743911
Page: 230

2. **Carson Kan**
Integrated Product Design MSc
chitpro_ductdesign@yahoo.com.hk
07708 094920
Page: 225

3. **Joel Kemp**
Industrial Design & Technology BA
joel@jh-kemp.com
www.jh-kemp.com
07901 536833
Pages: 66, 215

4. **Lucy Kierans**
Industrial Design & Technology BA
lekierans@hotmail.co.uk
www.lucyellendesign.co.uk
07943 838038
Pages: 131, 160

5. **Mina Kim**
Design Strategy & Innovation MA
Page: 188

6. **Benjamin Kirk**
Design & Branding Strategy MA
contact@benjaminkirk.com
www.benjaminkirk.com
07810 514150
Pages: 171 188

7. **Christopher Kneller**
Motorsport Engineering MEng
chris_kneller@hotmail.co.uk
07752 307956
Page: 297

8. **Chris Knight**
Mechanical Engineering MEng
cknight8@talktalk.net
07817 307354
Page: 306

9. **Emanuel Köchert**
Industrial Design & Technology BA
ekoechert@gmail.com
www.ekoechert.com
07510 344 112
Pages: 65, 269

10. **Katy Koren**
Product Design BSc
katy.koren@gmail.com
07827 668102
Pages: 72, 157

11. **Sze Yin Kwok**
Integrated Product Design MSc
kwokyin7@gmail.com
07403 862892
Page: 225

12. **Max Latimer**
Industrial Design & Technology BA
maxlatimer@hotmail.co.uk
www.maxlatimer.co.uk
07508 850954
Pages: 37, 132

13. **Hyein Lee**
Design & Branding Strategy MA
hilee0214@gmail.com
07714 202214
Pages: 170, 188

14. **Jiyun Lee**
Design & Branding Strategy MA
Jiyuen1113@gmail.com
07507 579301
Page: 241

15. **Sun Lee**
Design Strategy & Innovation MA
Page: 188

16. **Jermaine Legg**
Industrial Design & Technology BA
jermainelegg@gmail.com
07734 383781
Pages: 161, 223

1. **Zheng Li**
    Design & Branding Strategy MA
    abeeofchina@hotmail.com
    blog.sina.com.cn/matridom
    07586 725 715
    Page: 290

2. **Yanan Li**
    Design & Branding Strategy MA
    yananli.ice@gmail.com
    07872 492536
    Page: 192

3. **Yu-Ching Liao**
    Design Strategy & Innovation MA
    Page: 241

4. **Allan Lowther**
    Industrial Design & Technology BA
    allorus@allorus.com
    www.allorus.com
    07964 341819
    Page: 273

5. **Zhen Lun**
    Product Design BSc
    Lunzhen712@126.com
    07540 469693
    Page: 82

6. **Roman Luyken**
    Product Design Engineering BSc
    rl@studio-l.de
    www.studio-l.de
    +49 8151 79755
    Pages: 111, 268, 340

7. **Hugh Lyons**
    Integrated Product Design MSc
    contact@hughlyons.com
    www.hughlyons.com
    07709 903295
    Pages: 342, 345

8. **Kirsti Macqueen**
    Industrial Design & Technology BA
    kirsti.macqueen@gmail.com
    www.kirstimacqueendesign.co.uk
    07846 813860
    Pages: 24, 31

9. **Pritesh Makwana**
    Mechanical Engineering MEng
    pritesh_makwana@hotmail.com
    07847 320738
    Page: 313

10. **Rohan Malhotra**
    Integrated Product Design MSc
    rohan.2132680225@gmail.com
    07584 094606
    Pages: 270, 342

11. **Fawad Malik**
    Mechanical Engineering BEng
    Page: 309

12. **Michael Malone**
    Broadcast Media Design & Techology BSc
    mmalone89@googlemail.com
    07854 855357
    Page: 336

13. **Thomas Maltby**
    Product Design BSc
    temaltby@googlemail.com
    www.temaltby.com
    Pages: 267, 281

14. **Maria Paula Martinez Rodriguez**
    Design & Branding Strategy MA
    papaluskita@gmail.com
    coroflot.com/mariapaulamartinezr
    07576 424669
    Pages: 177, 194

15. **Rebecca McAvoy**
    Civil Engineering with
    Sustainability MEng
    Page: 214

16. **Samuel McClellan**
    Industrial Design & Technology BA
    samjmcclellan@gmail.com
    www.sammcclellan.co.uk
    07809 411138
    Pages: 45, 239

Mc - Pa

1. **Jennifer McCormack**
Design & Branding Strategy MA
jenny.mccormack@gmail.com
07906 848501
Pages: 173, 192

5. **Daniel Millard**
Mechanical Engineering with
Automotive Design MEng
dan.millard@hotmail.co.uk
07725 077363
Page: 296

9. **Jude Mudannayake**
Aviation Engineering BEng
jaam8@hotmail.com
07526 787547
Pages: 313

13. **Buster Palmano**
Product Design Engineering BSc
buster@palmano.co.uk
www.palmano.co.uk
07500 844559
Pages: 220, 248

2. **Yupeng Meng**
Integrated Product Design MSc
mengyupeng@126.com
07587 956074
Page: 231

6. **Premal Mistry**
Integrated Product Design MSc
premal_mistry@me.com
turbochargersonline.co.uk/p-mistry
07951 199951
Page: 156

10. **Stephen Nicholls**
Product Design Engineering BSc
steveo1@btopenworld.com
07796 170771
Page: 42

14. **Joseph Palmer**
Mechanical Engineering with
Aeronautics BEng
onetrickhorse@hotmail.com
07956 304938
Page: 229

3. **Efua Mensah-Ansong**
Industrial Design & Technology BA
efmens@hotmail.co.uk
www.efuadesign.com
07877 228042
Pages: 99, 243

7. **Unji Moir**
Product Design BSc
unjimoir@gmail.com
www.umoir.co.uk
07956 426424
Page: 233

11. **Rohin Odell**
Product Design Engineering BSc
rohin.odell@gmail.com
07984 227434
Pages: 49, 200

15. **Haeyoon Park**
Integrated Product Design MSc
haeyoonPark@hotmail.com
07538 511209
Pages: 193, 210

4. **Joe Midgley**
Industrial Design & Technology BA
jcl.midgley@gmail.com
www.jmid.co.uk
07790 837310
Pages: 27, 216, 238

8. **Nicholas Monk**
Aerospace Engineering MEng
nickmonk10@hotmail.co.uk
07969 007856
Page: 312

12. **Andra Oprisan**
Design & Branding Strategy MA
hello@andraoprisan.com
www.andraoprisan.com
07708 239207
Pages: 174, 192

16. **Matthew Parrish**
Product Design BSc
info@matthewparrishdesign.co.uk
www.matthewparrishdesign.com
07954 349058
Pages: 95, 201

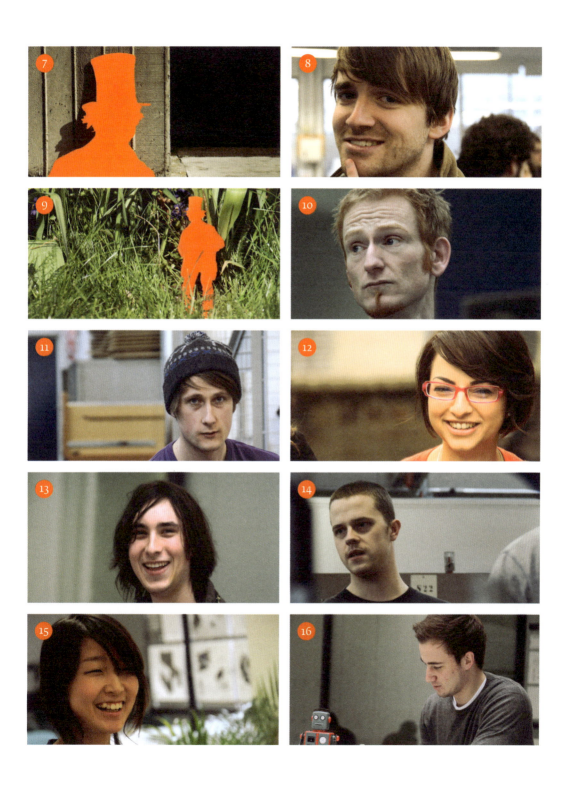

1. Rishee Patel
Multimedia Technology & Design BSc
rishee.patel@gmail.com
www.risheepatel.com
07545 472959
Page: 330

2. Adenike Pearse
Design & Branding Strategy MA
nikepeach@yahoo.com
07586 726056
Page: 191

3. Charlotte Pharez
Product Design Engineering BSc
charlotte_pharez@hotmail.co.uk
www.charlottepharez.co.uk
07821 770394
Pages: 43, 124

4. Antony Piper
Mechanical Engineering MEng
piperantony@hotmail.com
07554 665052
Page: 33

5. Michelle Poon
Product Design BSc
michelle_poon89@hotmail.com
07825 370375
Pages: 80, 113

6. Michael Puttock
Industrial Design & Technology BA
michael@mputtock.co.uk
www.mputtock.co.uk
07969 346188
Pages: 86, 289

7. Keahn Rahimi
Broadcast Media
Design & Techology BSc
keahnr@gmail.com
www.vimeo.com/keahn
07816 150583
Page: 334

8. Mohammed Rahman
Aerospace Engineering MSC
Page: 228

9. Anthony Ramnath
Multimedia Technology & Design BSc
anthony.ramnath@gmail.com
www.anthonyramnath.com
07754 955884
Page: 324

10. Tom Reader
Product Design BSc
tomreader88@hotmail.co.uk
07766 061461
Pages: 126, 293

11. Javad Rezaei Zadeh
Aerospace Engineering BEng
Page: 316

12. Reza Rezaei Zadeh
Aerospace Engineering BEng
Page: 316

13. Carlos Rojas Monserratte
Design & Branding Strategy MA
carlos.rojasm@gmail.com
07980 488100
Page: 191

14. Rebeckah Rose
Industrial Design & Technology BA
rebeckahlaureen@gmail.com
07875 686371
Page: 283

15. Matthew Rowinski
Industrial Design & Technology BA
m_rowinski@hotmail.com
www.matthewrowinski.com
07783 962703
Pages: 87, 287

16. Jonathan Rugg
Industrial Design & Technology BA
jontyrugg@msn.com
www.jontyrugg.com
07796 015715
Pages: 89, 277

1. Theo Rutter
    Mechanical Engineering MEng
    trutter123@MSN.com
    07789 310 870
    Page: 296

2. Calum Ryan
    Multimedia Technology & Design BSc
    calumeryan@gmail.com
    www.zizzfusion.com
    07910 124665
    Page: 333

3. Shabaz Sadiq
    Aerospace Engineering MSC
    Page: 228

4. Mohammed Umair
    Saeed Qureshi
    Mechanical Engineering BEng
    umair_6@hotmail.com
    07540 838575
    Page: 309

5. Tor Sandén
    Integrated Product Design MSc
    tor.sanden@gmail.com
    07572 128155
    Page: 211

6. Nick Sardar
    Product Design BSc
    hello@nicksardar.com
    www.nicksardar.com
    Twitter @NickSardar
    07792 827709
    Pages: 28, 73

7. Jenny Schneider
    Product Design BSc
    jennystardesign@hotmail.co.uk
    020 86563093
    Pages: 53, 64

8. George Schofield
    Aerospace Engineering MEng
    surebruv@hotmail.com
    07933 433849
    Page: 308

9. Dominic Sebastian
    Industrial Design & Technology BA
    dominicsebastian@live.co.uk
    www.dominicsebastian.com
    07825 322153
    Pages: 30, 41

10. Sathiya Sekar
    Aerospace Engineering MSC
    Page: 228

11. Arber Shabani
    Integrated Product Design MSc
    arber.005@gmail.com
    07999 469500
    Pages: 156, 342

12. Thomas Shadbolt
    Mechanical Engineering with
    Aeronautics BEng
    shadbolt_tom@hotmail.com
    07590 641516
    Page: 229

13. James Shakespeare
    Multimedia Technology & Design BSc
    j@jshakespeare.com
    www.jshakespeare.com
    07756 166366
    Page: 331

14. Tobias Shanker
    Mechanical Engineering MEng
    tobiasshanker@gmail.com
    02083 850216
    Page: 33

15. Hyunhee Shim
    Design & Branding Strategy MA
    shhhihi@never.com
    07411 897018
    Page: 189

16. Inderjit Sidhu
    Civil Engineering with
    Sustainability MEng
    Page: 214

1. **Michal Simko**
Mechanical Engineering with
Automotive Design MEng
michal.simko@hotmail.co.uk
07783 415247
Page: 305

2. **James Simpson**
Mechanical Engineering with
Aeronautics BEng
james.stephen.simpson@gmail.com
07840 144280
Page: 229

3. **Anita Singh**
Civil Engineering with
Sustainability MEng
Page: 214

4. **Adam Smith**
Industrial Design & Technology BA
design@adampaulsmith.co.uk
www.adampaulsmith.co.uk
07809 483858
Pages: 48, 74

5. **Daniel Smith**
Industrial Design & Technology BA
dan@smithdesign.eu
www.smithdesign.eu
07834 423069
Pages: 130, 227

6. **Jonathan Smith**
Aerospace Engineering BEng
jsmith29_9@msn.com
07787 142876
Page: 309

7. **Matt Smith**
Product Design BSc
smith89m@gmail.com
07890 956054
Pages: 90, 109

8. **Eunchung Song**
Design & Branding Strategy MA
dreamaker100@gmail.com
07810 861239
Page: 189

9. **Jung Sook Han**
Design & Branding Strategy MA
h1024415@gmail.com
www.koondesign.com
07412 009173
Page: 192

10. **Michael Stanley**
Industrial Design & Technology BA
hello@look-again.net
www.look-again.net
07872 024214
Pages: 205, 348

11. **Christopher Strickland**
Product Design Engineering BSc
hello@chrisstrickland.co.uk
www.chrisstrickland.co.uk
07722 055960
Pages: 22, 102, 251

12. **Saravana Subramony**
Aerospace Engineering MSC
Page: 228

13. **Sarp Suerdas**
Integrated Product Design MSc
ssuerdas@gmail.com
www.sarpsuerdas.net
07583 519733
Page: 344

14. **Min Suh**
Design & Branding Strategy MA
Pages: 193, 194

15. **Fred Swallow**
Industrial Design & Technology BA
fred.swallow@gmail.com
07590 836561
Pages: 163, 202

16. **Trevor Swayne**
Multimedia Technology & Design BSc
trevor.swayne@gmail.com
www.trevorswayne.co.uk
07895 513502
Page: 332

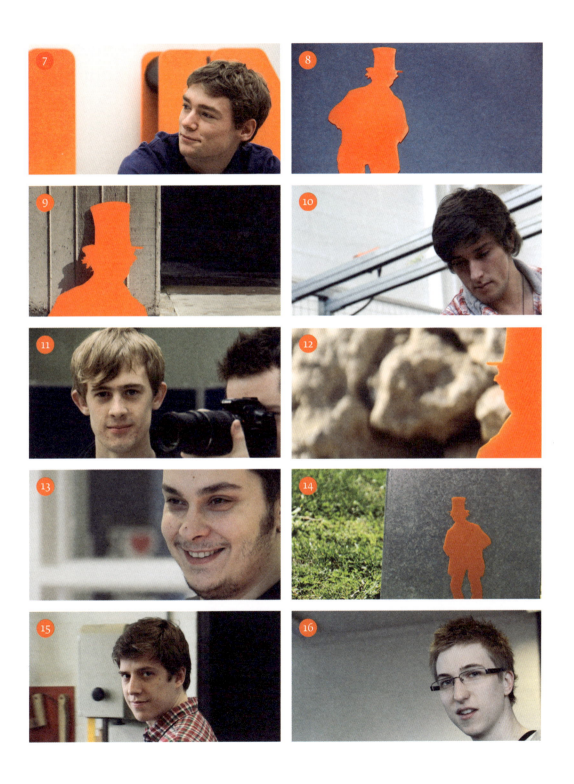

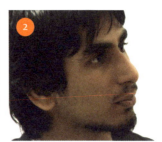

1. **Bernard Tashie-Lewis**
Aerospace Engineering MEng
ben4real2004@msn.com
07908 900307
Page: 317

2. **Ashwin Thirumurthy**
Integrated Product Design MSc
ashwin.t@live.com
07857 137536
Pages: 342, 343

3. **Philip Thomas**
Product Design BSc
contact@ptdesigns.co.uk
www.ptdesigns.co.uk
07968 630648
Pages: 203, 252

4. **Ross Thompson**
Product Design Engineering BSc
ross_e_thompson@hotmail.com
07919 433531
Pages: 294

5. **Toby Thwaites**
Aerospace Engineering MEng
toby.thwaites@yahoo.co.uk
07590 515867
Page: 312

6. **Simon Trevena**
Aerospace Engineering BEng
simon.trevena@googlemail.com
07834 584719
Page: 309

7. **Jamie Trigg**
Industrial Design & Technology BA
jamie.trigg@jtriggdesign.co.uk
www.jtriggdesign.co.uk
07985 121723
Pages: 91, 151

8. **Konstantina Trigkaki**
Design & Branding Strategy MA
nadiatrigaki@hotmail.com
07508 300985
Page: 189

9. **Ming-Chih Tsai**
Integrated Product Design MSc
Duewuo@gmail.com
07402 089581
Page: 210

10. **Theodore Tsikolis**
Design & Branding Strategy MA
teotsik@gmail.com
07526 011165
Pages: 178, 189, 194

11. **Jayson Tulloch**
Product Design BSc
jdtdesign@ymail.com
07817 142908
Pages: 141, 299

12. **Emma Tuttlebury**
Product Design BSc
hello@emmatuttlebury.co.uk
www.emmatuttlebury.co.uk
07941 564032
Pages: 107, 271

13. **Dimitra Tzarela**
Design & Branding Strategy MA
d.tzarela@yahoo.gr
07738 071012
Page: 189

14. **Tilia Van Olmen**
Integrated Product Design MSc
tiliavanolmen@googlemail.com
07722 825920
Page: 167

15. **Sam Verma**
Product Design BSc
sverma729@gmail.com
07913 778648
Pages: 150, 207

16. **Tom Wade**
Integrated Product Design MSc
tmrhwade@gmail.com
uk.linkedin.com/in/tomwadedesign
07775 910708
Page: 211

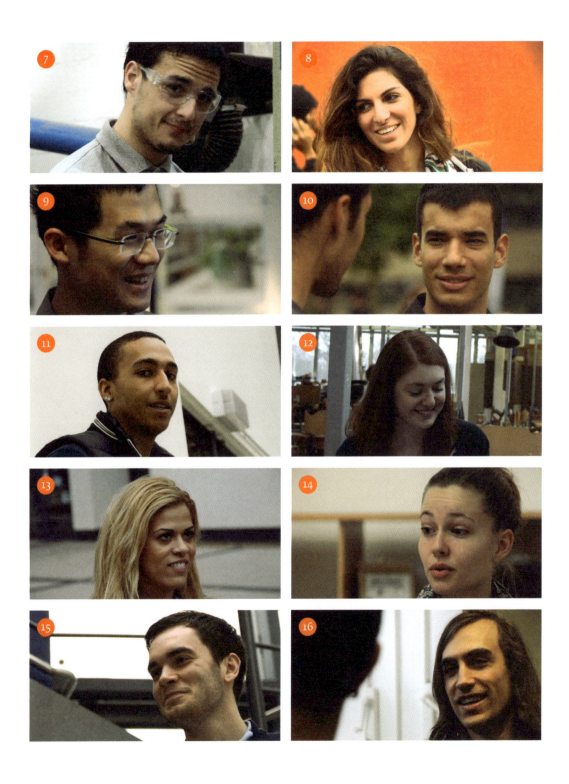

1. **Bing Wang**
Industrial Design & Technology BA
daisy.728@hotmail.com
07969 039938
Pages: 81, 115

2. **Matthew Watts**
Design & Branding Strategy MA
matthew@creativefuel.com.hk
07771 357686
Page: 179

3. **Marshel Weerakone**
Civil Enginering with
Sustainability MEng
weerakone@gmail.com
07515 124389
Page: 314

4. **Rosanna Wells**
Product Design Engineering BSc
rosiewells@hotmail.co.uk
www.rosannawells.co.uk
07841 424110
Pages: 32, 125

5. **Bradley Wherry**
Industrial Design & Technology BA
wherrydesign@gmail.com
www.wherrydesign.co.uk
07403 287850
Pages: 166, 240

6. **Adam White**
Mechanical Engineering MEng
adam.j.white1987@gmail.com
07906 311264
Page: 297

7. **Benjamin Whitehead**
Product Design BSc
hello@bwdesign.me.uk
www.bwdesign.me.uk
07886 892541
Pages: 106, 114

8. **Stuart Wickens**
Industrial Design & Technology BA
stuart@sawdesign.me.uk
www.sawdesign.me.uk
07955 818618
Pages: 133, 246

9. **Daniel Wilkin**
Industrial Design & Technology BA
danfwilkin@hotmail.co.uk
07753 345129
Page: 276

10. **Harrison Williams**
Product Design BSc
hw@ausendesign.com
www.ausendesign.com
07590 587019
Pages: 221, 250, 328

11. **Peter Williams**
Product Design BSc
peter@peterwilliamsdesign.co.uk
www.peterwilliamsdesign.co.uk
07933 543810
Pages: 36, 222

12. **Alex Williamson**
Product Design Engineering BSc
alex.williamson@xiidesign.com
www.xiidesign.com
07796 177289
Pages: 93, 149

13. **James Willson**
Industrial Design & Technology BA
james_willson@hotmail.com
www.i3design.com
07515 558332
Pages: 88, 247

14. **Samuel Wong**
Product Design Engineering BSc
scswong@hotmail.co.uk
07967 383930
Pages: 38, 46, 77

15. **Sunhye Woo**
Design & Branding Strategy MA
Page: 193

16. **Rob Wooldridge**
Industrial Design & Technology BA
vonwolfindustries@hotmail.co.uk
07919 251449
Pages: 134

1. **Max Woźniak**
Product Design BSc
wozmv88@gmail.com
www.angryswans.com
07731 470250
Pages: 15, 75, 291

2. **Qiqi Xiang**
Design & Branding Strategy MA
xiangqiqi@yahoo.com.cn
07759 326889
Page: 190

3. **Tingting Xiao**
Integrated Product Design MSc
alittlett@126.com
07809 761824
Page: 52

4. **Haoyun Xue**
Integrated Product Design MSc
xuehaoyun@gmail.com
07586 725766
Page: 345

5. **Meihui Yan**
Industrial Design & Technology BA
maice_yan@hotmail.com
Page: 83

6. **Ivan Yang**
Integrated Product Design MSc
ivan92953@gmail.com
07856 441535
Pages: 194, 230

7. **Shuzhen Yang**
Design & Branding Strategy MA
laura.oo.yang@gmail.com
07586 725800
Page: 190

8. **Wei Ye**
Design & Branding Strategy MA
willye509@gmail.com
07760 679001
Page: 190

9. **Charles Youell**
Mechanical Engineering MEng
charles.y87@gmail.com
07988 899 085
Page: 296

10. **Chi Yu**
Multimedia Technology & Design BSc
chi@chiwayu.co.uk
www.chiwayu.co.uk
07515 104773
Page: 337

11. **Xiaofan Yu**
Design & Branding Strategy MA
www.xiaofanyu.com
07773 296887
Page: 190

12. **Johanna Zambrano Novoa**
Design & Branding Strategy MA
johannazam@gmail.com
07735 351828
Page: 189

13. **Philip Zeitler**
Industrial Design & Technology BA
philip_zeitler@msn.com
www.pzeitler-design.de
(+49)173 593 7515
Pages: 57, 278

14. **Hongchen Zhang**
Integrated Product Design MSc
lyndonzhang1985@gmail.com
07403 300695
Page: 181

15. **Lu Zhang**
Design & Branding Strategy MA
angelaz7777@gmail.com
07587 325540
Page: 232

People. What would we do without you?

One thing that often manages to escape these pages is the fact that Made in Brunel only exists because many people contribute their spare time to make it happen. It's a massive undertaking and responsibility, even if we had all the time in the world.

But, so many of you are willing to give up what little time you have. You balance your careers with an extra workload. You keep your workshops open late. You stay on your feet for days sawing, drilling and painting. You work into the small hours proofreading. You lend us equipment to make our book. You go out of your way to help us…

Every year, all of you add to the heart and soul that pours into Made in Brunel. If this book was big enough to represent our gratitude, the world would sit nicely inside an "o".

Thank you.

Made in Brunel